Encyclopedia Brittanica

Encyclopedia Britannica : Microscope

Vol IX

Encyclopedia Brittanica

Encyclopedia Britannica : Microscope
Vol IX

ISBN/EAN: 9783742823519

Manufactured in Europe, USA, Canada, Australia, Japa

Cover: Foto ©Andreas Hilbeck / pixelio.de

Manufactured and distributed by brebook publishing software
(www.brebook.com)

Encyclopedia Brittanica

Encyclopedia Britannica : Microscope

ENCYCLOPÆDIA BRITANNICA;

OR, A

DICTIONARY

OF

ARTS, SCIENCES,

AND

MISCELLANEOUS LITERATURE;

Conſtructed on a PLAN,

BY WHICH

THE DIFFERENT SCIENCES AND ARTS

Are digeſted into the FORM of Diſtinct

TREATISES OR SYSTEMS,

COMPREHENDING

The HISTORY, THEORY, and PRACTICE, of each,
according to the Lateſt Diſcoveries and Improvements;

AND FULL *EXPLANATIONS* GIVEN OF THE

VARIOUS DETACHED PARTS OF KNOWLEDGE,

WHETHER RELATING TO

NATURAL and ARTIFICIAL Objects, or to Matters ECCLESIASTICAL,
CIVIL, MILITARY, COMMERCIAL, &c.

Including ELUCIDATIONS of the moſt important Topics relative to RELIGION, MORALS,
MANNERS, and the OECONOMY of LIFE:

TOGETHER WITH

A DESCRIPTION of all the Countries, Cities, principal Mountains, Seas, Rivers, &c.
throughout the WORLD;

A General HISTORY, *Ancient* and *Modern,* of the different Empires, Kingdoms, and States;

AND

An Account of the LIVES of the moſt Eminent Perſons in every Nation,
from the earlieſt ages down to the preſent times.

Compiled from the writings of the beſt Authors, in ſeveral languages; the moſt approved Dictionaries, as well of general ſcience as of its particular branches; the Tranſactions, Journals, and Memoirs, of Learned Societies, both at home and abroad; the MS. Lectures of Eminent Profeſſors on different ſciences; and a variety of Original Materials, furniſhed by an Extenſive Correſpondence.

THE THIRD EDITION, IN EIGHTEEN VOLUMES, GREATLY IMPROVED.

ILLUSTRATED WITH FIVE HUNDRED AND FORTY-TWO COPPERPLATES.

VOL. XI.

INDOCTI DISCANT, ET AMENT MEMINISSE PERITI.

EDINBURGH:

PRINTED FOR A. BELL AND C. MACFARQUHAR.

M.DCC.XCVII.

Mick'e

Spenser," 4to; and in 1769 he published, " A Letter to Mr Harwood, wherein some of his creative Glosses, false Translations, and blundering Criticisms, in support of the Arian Heresy, contained in his Literal translation of the New Testament, are pointed out and confuted,' &c. and next year he published " Mary Queen of Scots, an Elegy ;" " Hengist and Mary, a Ballad;" and " Knowledge, an Ode ;" in Pearch's Collection of Poems. In 1772 he published " Voltaire in the Shades, or Dialogues on the Deistical Controversy," 8vo. The Elegy on Mary had been submitted to the judgment of Lord Lyttelton, who declined to criticise it, not for its deficiency in poetical merit, but from thinking differently from the author concerning that unfortunate princess.

About this time Mr Mickle was a frequent writer in the Whitehall Evening Post; but a more important work now engaged his attention. When no more than 17 years of age he had read Castera's translation of the Lusiad of Camoens into French, and then projected the design of giving an English translation of it. From this, however, he was prevented by various avocations till the year 1771, when he published the first book as a specimen ; and having prepared himself by acquiring some knowledge of the Portuguese language, he determined to apply himself entirely to this work. With this view he quitted his residence at Oxford, and went to a farm-house at Forest-hill, where he pursued his design with unremitting assiduity till the year 1775, when the work was entirely finished.

During the time that Mr Mickle was engaged in this work, he subsisted entirely by his employment as corrector of the press; and on his quitting that employment he had only the subscriptions he received for his translation to support him. Notwithstanding these difficulties, he adhered steadily to the plan he had laid down, and completed it in about five years.

When his work was finished, Mr Mickle applied to a person of great rank, with whom his family had been connected, for permission to dedicate it to him. Permission was granted, and his patron honoured him with a very polite letter; but after receiving a copy, for which an extraordinary price was paid for the binding, he did not think proper to take any notice of the author. At last a gentleman of high rank in the political world, a firm friend to the author, and who afterwards took him under his protection, waited on the patron, and heard him declare that he had not read the work, but that it had been reprobated not to have the merit it was at first said to possess. The applause with which the work was received, however, soon banished from the author's mind those disagreeable sensations which had been occasioned by the contemptuous neglect of his patron, as well as those severe criticisms which had been circulated concerning it. A second edition was prepared in 1778, with a plate prefixed to it, executed by the celebrated artist Mortimer; on whom Mr Mickle wrote an epitaph in 1779. This year also he published a pamphlet, intitled, " A Candid Examination of the Reasons for depriving the East India Company of its Charter, contained in The History and Management of the East India Company from its Commencement to the Present Time; together with some Strictures on the Self-Contradictions and

Historical Errors of Dr Adam Smith, in his Reasons for the Abolition of the said Company," 4to. About this time some of his friends thought of recommending him to the king as deserving of a pension ; but this scheme was never put in execution. Dr Lowth, bishop of London, would have put him into orders, and provided for him in the church; but this was not agreeable to our author's disposition. While he was meditating a publication of all his poems, in which he would most probably have found his account, he was appointed secretary to Commodore Johnstone, who had lately obtained the command of the Romney man of war. In November 1779 he arrived at Lisbon, and was named by his patron joint agent for the prizes which were taken. In this capital and its neighbourhood he resided more than six months, being every where received with every mark of politeness and attention ; and during this period he composed his poem called " Almada Hill," which in 1781 was published in quarto. He collected also many particulars concerning the manners of the Portuguese, which he intended also to have published. During his stay at Lisbon the royal academy was opened; and Mr Mickle, who was present at the ceremony of its commencement, had the honour to be admitted a member under the presidency of Don John of Braganza duke of Lafoens. His presence being thought necessary in England to attend to the proceedings of the courts of law respecting the condemnation of some of the prizes, he did not accompany the commodore in his last expedition, nor did he go any more to sea. In 1782 he published " The Prophecy of Queen Emma, an ancient Ballad lately discovered, written by Johannes Turgottus, prior of Durham, in the reign of William Rufus; to which is added by the Editor, an Account of the Discovery, and Hints towards a Vindication of the Authenticity of the Poems of Ossian and Rowley," 8vo.

In June this year Mr Mickle married Miss Tomkins, daughter of the person with whom he resided at Forest-hill, while engaged in translating the Lusiad. Having received some fortune with this lady, as well as made some money himself when in the service of Commodore Johnstone, he now enjoyed a comfortable independence. Having fixed his residence at Wheatley in Oxfordshire, he devoted his time to the revision of his poetical works, which he proposed to publish by subscription; but the plan has not yet been carried into execution. The last seven years of his life were employed in writing for the European Magazine. The Fragments of Leo, and some of the most approved reviews of books, in that performance, were of his production. He died after a short illness on the 25th of October 1788 at Wheatley, leaving one son behind him. His poetry possesses much beauty, variety, harmony of numbers, and vigour of imagination; his life was without reproach; his foibles were few; and his offences; his virtues many; and his genius very considerable.

MICROCOSM, a Greek term signifying the little world; used by some for man, as being supposed an epitome of the universe or great world.

MICROCOSMIC acid. See Phosphorus (Acid of).

MICRO-

MICROGRAPHY, the description of objects too minute to be viewed without the affistance of a microscope. See *Microscopic Objects*.

MICROMETER, an instrument, by the help of which the apparent magnitudes of objects viewed thro' telescopes or microscopes are measured with great exactness.

I. The first telescopic micrometers were only mechanical contrivances for measuring the image of an object in the focus of the object-glass. Before these contrivances were thought of, astronomers were accustomed to measure the field of view in each of their telescopes, by observing how much of the moon they could see through it, the semidiameter being reckoned at 15 or 16 minutes; and other distances were estimated by the eye, comparing them with the field of view. Mr Galcoigne, an English gentleman, however, fell upon a much more exact method, and had a Treatise on Optics prepared for the press; but he was killed during the civil wars in the service of Charles I. and his manuscript was never found. His instrument, however, fell into the hands of Mr R. Townly, who fays, that by the help of it he could mark above 40,000 divisions in a foot.

Mr Galcoigne's instrument being shown to Dr Hooke, he gave a drawing and description of it, and proposed several improvements in it, which may be seen in Phil. Trans. abr. Vol. I. p. 217. Mr Galcoigne divided the image of an object, in the focus of the object-glass, by the approach of two pieces of metal ground to a very fine edge, in the place of which Dr Hooke would substitute two fine hairs stretched parallel to one another. Two other methods of Dr Hooke's, different from this, are described in his Posthumous Works, p. 497, 498. An account of several curious observations that Mr Galcoigne made by the help of his micrometer, particularly in the mensuration of the diameters of the moon and other planets, may be seen in the Phil. Trans. Vol. XLVIII. p. 190.

Mr Huygens, as appears by his System of Saturn, published in 1659, used to measure the apparent diameters of the planets, or any small angles, by first measuring the quantity of the field of view in his telescope; which, he fays, is best done by observing the time which a star takes up in passing over it, and then preparing two or three long and slender brass plates, of various breadths, the sides of which were very straight, and converging to a small angle. In making use of these pieces of brass, he made them slide in two slits, that were made in the sides of the tube, opposite to the place of the image, and observed in what place it just covered the diameter of any planet, or any small distance that he wanted to measure. It was observed, however, by Sir Isaac Newton, that the diameters of planets, measured in this manner, will be larger than they should be, as all lucid objects appear to be when they are viewed upon dark ones.

In the Ephemerides of the Marquis of Malvasia, published in 1662, it appears that he had a method of measuring small distances between fixed stars and the diameters of the planets, and also of taking accurate draughts of the spots of the moon; and this was by a net of silver wire, fixed in the common focus of the object and eye-glass. He also contrived to make one of two stars to pass along the threads of this net, by turning it, or the telescope, as much as was necessary for

that purpose; and he counted, by a pendulum-clock, beating seconds, the time that elapsed in its passage from one wire to another, which gave him the number of the minutes and feconds of a degree contained between the intervals of the wires of his net, with respect to the focal length of his telescope.

In 1666, Messrs Azout and Picard published a description of a micrometer, which was nearly the same with that of the Marquis of Malvasia, excepting the method of dividing it, which they performed with more exactness by a screw. In some cases they used threads of silk, as being finer than silver wires. Decbales also recommends a micrometer consisting of fine wires, or silken threads, the distances of which were exactly known, disposed in the form of a net, as peculiarly convenient for taking a map of the moon.

M. de la Hire fays, that there is no method more simple or commodious for observing the digits of an eclipse than a net in the focus of the telescope. These, he fays, were generally made of silken threads; and that for this particular purpose six concentric circles had also been made use of, drawn upon oiled paper; but he advises to draw the circles on very thin pieces of glass with the point of a diamond. He also gives several particular directions to assist persons in the use of them. In another memoir he shows a method of making use of the same net for all eclipses, by using a telescope with two object-glasses, and placing them at different distances from one another.

Different Constructions of Micrometers. The first we shall describe is that by Mr Huygens. Let ABCD be a section of the telescope at the principal focus of the object-glass, or where the wires are situated, which are placed in a sheet tube containing the eye-glass, and may be turned into any position by turning that tube; m a is a fine wire extended over its centre; o a, x y, are two straight plates whose edges are parallel and well defined, and perpendicular to m a; o a is fixed, and x y moves parallel to it by means of a screw, which carries two indexes over a graduated plate, to show the number of revolutions and parts of a revolution which it makes. Now to measure any angle, we must first ascertain the number of revolutions and parts of a revolution corresponding to some known angle, which may be thus done. 1st, Bring the inner edges of the plates exactly to coincide, and set each index to o; turn the screw, and separate the plates to any distance; and observe the time a star m i, in passing along the wire m a from one plate to the other: for that time, turned into minutes and feconds of a degree, will be the angle answering to the number of revolutions, or the angle corresponding to the distance. Thus, if d m cos. of the star's declination, we have 15' d m, the angle corresponding to this distance; and hence, by proportion, we find the angle answering to any other. 2dly, Set up an object of a known diameter, or two objects at a given distance, and turn the screw till the edges of the plates become tangents to the object, or till their opening just takes in the distance of the two objects upon the wire m a; then from the diameter, or distance of the two objects from each other, and their distance from the glass, calculate the angle, and observe the number of revolutions and parts corresponding. 3dly, Take the diameter of the sun on any day, by making the edges of the plates tangents to the opposite limbs, and find, from

Plate CCCCXCV.
fig. 1.

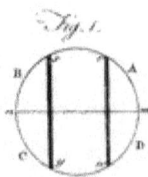

Fig. 1.

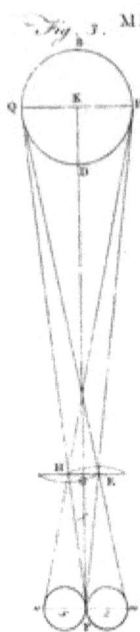

Fig. 3.

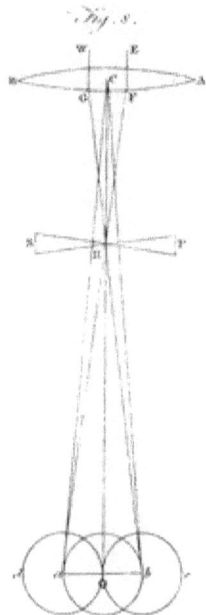

Fig. 8.

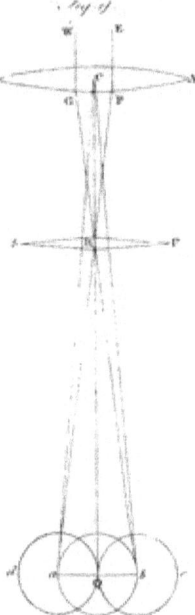

Fig. 9.

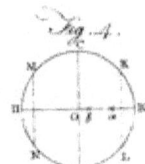

Fig. 4.

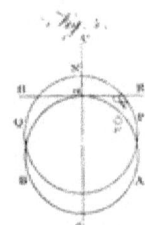

Fig. 5.

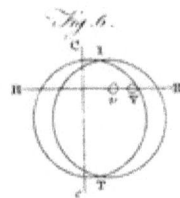

Fig. 6.

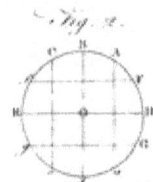

Fig. 2.

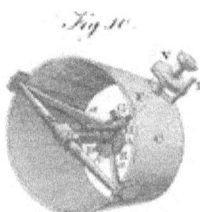

Fig. 10.

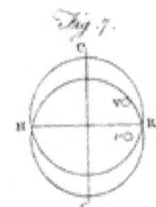

Fig. 7.

Fig. 11.

Microme-
ter

from the nautical almanac, what is his diameter on
that day. Here it will be best to take the upper and
lower limbs of the sun when on the meridian, as he has
then no motion perpendicular to the horizon. If the
edges do not coincide when the indexes stand at ●,
we must allow for the error. Instead of making a proportion, it is better to have a table calculated to show
the angle corresponding to every revolution and parts
of a revolution. But the observer must remember, that
when the micrometer is fixed to telescopes of different
focal lengths, a new table must be made. The whole
system of wires is turned about in its own plane, by
turning the eye-tube round with a hand, and by
that means the wire ● ● can be thrown into any position, and consequently angles in any position may be
measured. Dr Bradley added a small motion by a
rack and pinion to set the wires more accurately in any
position.

Instead of two plates, two wires were afterwards
put; and Sir Isaac Newton observed, that the diameters of the planets measured by the plates were somewhat bigger than they ought, as appeared by comparing Mr Huygens's measures with others taken with the
wires; and also by comparing the diameter of mercury observed in and out of the sun's disk, the latter
being the greatest. Dark objects on bright ones appear less, and light objects on dark ones appear greater,
than if they were equally bright; owing, perhaps, to
the brighter image on the retina diffusing itself into the
darker; and the bright image of the planet being in-
tercepted by the plates, the faint diffused light becomes
more sensible, and is mistaken for the edge of the planet.

But the micrometer, as now contrived, is of use,
not only to find the angular distance of bodies in
the field of view at the same time, but also of those
which, when the telescope is fixed, pass through the
field of view successively; by which means we can find
the difference of their right ascensions and declinations.
Let A a, B b, C c, be three parallel and equidistant
wires, the middle one bisecting the field of view;
H O R a fixed wire perpendicular to them passing
through the centre of the field; and F f, G g, two
wires parallel to it, each moveable by a micrometer screw, as before, so that they can be brought
up to H O R, or a little beyond. Then to find
the angular distance of two objects, bring them very
near to B●, and in a line parallel to it, by turning
about the wires, and bring one upon HOR, and by the
micrometer screw make F f or G g pass through the
other; then turn the screw till that wire coincides
with H O R, and the arc which the index has passed
over shows their angular distance. If the object be
further remote than you can carry the distance of one
of the wires F f, G g, from H O R, then bring one
object to F f and the other to G g; and turn each
micrometer screw till they meet, and the sum of the
arcs passed over by each index gives their angular distance. If the objects be two stars, and one of them
be made to run along H O R, or either of the moveable wires as occasion may require, the motion of the
other will be parallel to these wires, and their difference of declinations may be observed with great exactness; but in taking any other distances, the motion

of the stars being oblique to them, it is not quite so
easy to get them parallel to B f; because if one star
be brought near, and the eye be applied to the other
to adjust the wires to it, the former star will have got-
ten a little away from the wire. Dr Bradley, in his
account of the use of this micrometer, published by
Dr Mukelipne in the Philosophical Transactions
for 1772, thinks the best way is to move the eye backwards and forwards as quick as possible; but it seems
to me to be best to fix the eye at some point between,
by which means it takes in both at once sufficiently
well defined to compare them with B●. In finding
the difference of declinations, if both bodies do not
come into the field of view at the same time, make
one run along the wire H O R, as before, and fix the
telescope and wait till the other comes in, and then
adjust one of the moveable wires to it, and bring it up
to H O R, and the index gives the difference of their
declinations. The difference of time between the passage of the star at either of the two moveable wires,
and the transit of the other star over the cross fixed
wire (which represents a meridian), turned into degrees and minutes, will give the difference of right
ascension. The star has been here supposed to be bi-
sected by the wire; but if the wire be a tangent to it,
allowance must be made for the breadth of the wire,
provided the adjustment be made for the coincidence of the wires. In observing the diameters of the
sun, moon, or planets, it may perhaps be most convenient to make use of the outer edges of the wires,
because they appear most distinct when quite within
the limb; but if there should be any sensible reflection
of the rays of light in passing by the wires, it will be
best avoided by using the inner edge of one wire and
the outward edge of the other; for by that means
the inflection at both limbs will be the same way,
and therefore there will be no alteration of the relative position of the rays passing by each wire. And
it will be convenient in the micrometer to note at what
division the index stands when the moveable wire coincides with H O R; for then you need not bring the
wire when a star is upon it up to H O R, may reckon
from the division at which the index then stands to the
above division.

With a micrometer therefore thus adapted to a telescope, Mr Herrington Rowley of Exeter proposed
a new way of measuring the difference between the
greatest and least apparent diameters of the sun, although the whole of the sun was not visible in the field
of view at once. The method we shall briefly describe.
Since two object-glasses instead of one, so as to form
two images whose limbs shall be at a small distance
from each other; or instead of two perfect lenses, he
proposed to cut a single lens into four pairs of equal
breadths by parallel lines, and to place the two segments with their straight sides against each other, or
the two middle fourteens with their opposite edges
together; in either case, the two parts which before
had a common centre and axis, have now their centres
and axes separated, and consequently two images will
be formed as before by two perfect lenses. Another
method in reflectors was to cut the large concave reflector through the centre, and by a contrivance to
turn up the outer edges while the straight ones remain

Microme-
ter

Microme-
ter.

mained fixed; by which means the axis of the two
parts became inclined, and formed two images. Two
images being formed in this manner, he proposed to
measure the distance between the limbs when the diameters of the sun were the greatest and least, the difference of which would be the difference of the diameters required. Thus far we are indebted to Mr Sa-
very for the idea of forming two images; and the ad-
mirable uses to which it was afterwards applied, we
shall next proceed to describe.

The divided object-glass micrometer, as now made,
was contrived by the late Mr John Dollond, and by
him adapted to the object-end of a reflecting telescope,
and has been since by the present Mr P. Dollond his son
applied with equal advantage to the end of an achromatic
telescope. The principle is this: The object-glass is
divided into two segments in a line drawn through the
centre; each segment is fixed in a separate frame of
brass, which is moveable, so that the centres of the
two segments may be brought together by a handle
for that purpose, and thereby form one image of an
object; but when separated they will form two images, lying in a line passing through the centre of each
segment; and consequently the motion of each image
will be parallel to that line, which can be thrown into
any position by the contrivance of another handle to
turn the glass about in its own plane. The brass-work
carries a vernier to measure the distance of the centres
of the two segments. Now let E and H be the centres of the two segments, F their principal focus, and
P Q two distant objects in FE, FH, produced, or the
opposite limbs of the same object PBQD; then the
images of P and Q, formed by each segment, or the
images of the opposite limbs of the object PBQD,
coincide at F: hence two images as P, as F of that
object are formed, whose limbs are in contact; therefore the angular distance of the points P and Q is the
same as the angle which the distance EH subtends at
F, which, as the angles supposed to be measured are
very small, will vary as EH extremely nearly; and
consequently if the angle corresponding to one interval of the centres of the segments be known, the
angle corresponding to any other will be found by proportion. Now to find the interval for some one angle,

N° 248.

Fig. 3.

take the horizontal diameter of the sun on any day,
by separating the images till the contrary limbs coincide, and read off by the vernier the interval of their
centres, and look into the nautical almanac for the diameter of the sun on that day, and you have the corresponding angle. Or if greater exactness be required
than from taking the angle in proportion to the distances of their centres, we may proceed thus:—Draw
FG perpendicular to EH, which therefore bisects it;
then one half EH, or EG, is the tangent of half the
angle EFH; hence, half the distance of their centres
: : half any other distance of the centres : tangent of
half the corresponding angle (A).

Hence the method of measuring small angles is
manifest; for we consider P, Q, either as two objects whose images are brought together by separating
the two segments, or as the opposite limbs of one object PBQA, whose images, formed by the two segments E, H, touch at F: in the former case, EH
gives the angular distance of the two objects; and in
the latter, it gives the angle under which the diameter
of the object appears. Hence, to find the angular
distance of two objects, separate the segments till the
two images which approach (B) each other coincide;
and to find the diameter of an object, separate the
segments till the contrary limbs of the images touch
each other, and read off the distance of the centres
of the segment from the vernier (c), and find the
angle as directed in the last article. From hence
appears one great superiority in this above the wire
micrometer; as, with this, any diameter of an object may be measured with the same ease and accuracy; whereas with that we cannot with accuracy
measure any diameter, except that which is at right
angles to its apparent motion.

But, besides these two uses to which the instrument
seems so well adapted, Dr Maskelyne has shown, in
the Philosophical Transactions for the year 1771,
how it may be applied to find the difference of right
ascensions and declinations. For this purpose, two
wires at right angles to each other, bisecting the field
of view, must be placed in the principal focus of the
eye-glass, and moveable about in their own plane.—

Microme-
ter.

Let

(A) If the object be not a distant one, let *f* be the principal focus; then F*f* : FG : : FG : FK (FG
being produced to meet a line joining the apparent places of the two objects P, Q), ∴ dividendo, *f*G : FG
: : GK : FK, and alternando, *f*G : GK : : FG : FK : : (by similar triangles) EH : PQ, hence $\frac{EH}{fG} = \frac{PQ}{GK}$
therefore the angle subtended by EH at *f* = the angle subtended by PQ at G; and consequently, as *f*G is
constant, the angle measured at G is, in this case also, in proportion to EH. The instrument is not adapted
to measure the angular distance of bodies, one of which is near and the other at a distance, because their
images would not be formed together.

(B) Besides these two images, there will be two others receding from each other, for each segment gives
an image of each object.

(C) To determine whether there be any error of adjustment of the micrometer scale, measure the diameter of any small well defined object, as Jupiter's equatorial diameter, or the longest axis of Saturn's ring,
both ways, that is, with o on the vernier to the right and left of o on the scale, and half the difference
is the error required; which must be added to   subtracted from all observations, according as the diameter measured with o on the vernier, when advanced on the scale, is less or greater than the diameter
measured the other way. And it is also evident, that half the sum of the diameters thus measured gives
the true diameter of the object.

Micrometer.

Fig. 4.

Let HCR be the field of view, HR and Cr the two wires; turn the wires till the westernmost star (which is the best, having further to move) run along ROH; then separate the two segments and turn about the micrometer till the two images of the same star lie in the wire C r and then, partly by separating the segments, and partly by raising or depressing the telescope, bring the two innermost images of the two stars to appear and run along ROH, as o, A, and the vernier will give the difference of their declinations; because, as the two images of one of the stars coincided with C, the image of each star was brought perpendicularly upon HR, as in their proper meridian. And, for the same reason, the difference of their times of passing the wire CO will give their difference of right ascensions. These operations will be facilitated, if the telescope be mounted on a polar axis. If two other wires KL, MN, parallel to C, be placed near H and R, the observation may be made on two stars whose difference of meridians is nearly equal to HR the diameter of the field of view, by bringing the two images of one of the stars to coincide with one of these wires. If two stars be observed whose difference of declinations is well settled, the scale of the micrometer will be known.

It has hitherto been supposed, that the images of the two stars can be both brought into the field of view at once upon the wire HOR; but if they cannot, set the micrometer to the difference of their declinations as nearly as you can, and make the image which comes first run along the wire HOR, by elevating or depressing the telescope; and when the other star comes in, if it do not also run along HOR, alter the micrometer till it does, and half the sum of the numbers shown by the micrometer at the two separate observations of the two stars on the wire HOR will be the difference of their declinations. That this should be true, it is manifestly necessary that the two segments should recede equally in opposite directions; and this is effected by Mr Dollond in his new improvement of the object-glass micrometer.

Fig. 5.

The difference of right ascensions and declinations of Venus or Mercury in the sun's disk and the sun's limb may be thus found. Turn the wires so that the north limb n of the sun's image AB, or the north limb of the image V of the planet, may run along the wire RH, which therefore will then be parallel to the equator, and consequently C a secondary to it; then separate the segments, and turn about the micrometer till the two images Vn of the planet pass Cr at the same time, and then by separating the segments, bring the north limb of the northernmost image V of the planet to touch HR, at the time the northernmost limb n of the southernmost image AB of the sun touches it, and the micrometer shows the difference of declinations of the northernmost limbs of the planet and sun, for the reason formerly given †, we having brought the northernmost limbs of the two innermost images V and AB to HR, these two being manifestly interior to n and the southernmost limb N of the image PQ. In the same manner we take the difference of declinations of their southernmost limbs; and

† See the preceding † also col 1. par. 4.

Vol. XI. Part II.

half the difference of the two measures, (taking immediately one after another) is equal to the difference of the declinations of their centres, without any regard to the sun's or planet's diameters, or error of adjustment of the micrometer; for as it affects both equally, the difference is the same as if there were no error: and the difference of the times of the transits of the eastern or western limbs of the sun and planet over Cr gives the difference of their right ascensions.

Micrometer.

Instead of the difference of right ascensions, the distance of the planet from the sun's limb, in lines parallel to the equator, may be more accurately observed thus: Separate the segments, and turn about the wires and micrometer, so as to make both images V, u, run along HR, or so that the two intersections I, T, of the sun's image may pass Cr at the same time. Then bring the planet's and sun's limbs into contact, as at V, and do the same for the other limb of the sun, and half the difference gives the distance of the centre of the planet from the middle of the chord on the sun's disk parallel to the equator, or the difference of the right ascensions of their centres, allowing for the motion of the planet in the interval of the observations, without any regard to the error of adjustment, for the same reason as before. For if you take any point in the chord of a circle, half the difference of the two segments is manifestly the distance of the point from the middle of the chord; and as the planet runs along HR, the chord is parallel to the equator.

Fig. 6.

In like manner, the distances of their limbs may be measured in lines perpendicular to the equator, by bringing the micrometer into the position already described*, and instead of bringing V to HR, separate the segments till the northernmost limbs coincide as at V; and in the same manner make their southernmost images to coincide, and half the difference of the two measures, allowing for the planet's motion, gives the difference of the declinations of their centres.

Fig. 7.

* See the preceding par. 3.

Hence the true place of a planet in the sun's disk may at any time of its transit be found; and consequently the nearest approach to the centre and the time of ecliptic conjunction may be deduced, although the middle should not be observed.

But however valuable the object-glass micrometer undoubtedly is, difficulties sometimes have been found in its use, owing to the alteration of the focus of the eye, which will cause it to give different measures of the same angle at different times. For instance, in measuring the sun's diameter, the axis of the pencil coming through the two segments from the contrary limbs of the sun, as PF, QF, fig. 3. crossing one another in the focus F under an angle equal to the sun's semidiameter, the union of the limbs cannot appear perfect, unless the eye be disposed to see objects distinctly at the place where the images are formed; for if the eye be disposed to see objects nearer to or further off than this place, in the latter case the limbs will appear separated, and in the former they will appear to lap over (D). This imperfection led Dr Maskelyne

4 T

(D) For if the eye can see distinctly an image at F, the pencils of rays, of which PF, QF are the two axes, diverging from F, are each brought to a focus on the retina at the same point; and therefore the two limbs appear

Micrometer. skelyne to inquire, whether some method might not be found of producing two distinct images of the sun, or any other object, by bringing the axis of each pencil to coincide, or very nearly so, before the formation of the images, by which means the limbs when brought together would not be liable to appear separated from any alteration of the eye; and this he found would be effected by the refraction of two prisms, placed either without or within the telescope; and on this principle, placing the prisms within, he constructed a new micrometer, and had one executed by Mr Dollond, which upon trial answered as he expected. The construction is as follows.

Fig. 1, 2. Let AB be the object-glass; ab the image, suppose of the sun, which would have been formed in the principal focus Q; but let the prisms PR, SR be placed to intercept the rays, and let EF, WG, be two rays proceeding from the eastern and western limbs of the sun, converging, after refraction at the lens, to a and b; and suppose the refraction of the prisms to be such, that in fig. 8. the ray EFR, after refraction at R by the prism PR, may proceed in the direction RQ; and as all the rays which were proceeding to a suffer the same refraction at the prism, they will all be refracted to Q; and therefore, instead of an image ab, which would have been formed by the lens alone, an image Q is formed by those rays which fall on the prism PR; and for the same reason, the rays falling on the prism SR will form an image Q: and in fig. 9. the image of the point b is brought to Q, by the prism PR, and consequently an image Q is formed by those rays which fall on PR; and for the same reason, an image Q is formed by the rays falling on SR. Now in both cases, as the rays EFR, WGR, coming from the two opposite limbs of the sun, and forming the point of contact of the two limbs, proceed in the same direction RQ, they must thus accompany each other through the eye-glass and also through the eye, whatever refractive power it has, and therefore to every eye the images must appear to touch. Now the angle aRb is twice the refraction of the prism, and the angle aCb is the diameter of the sun; and as these angles are very small, and have the same subtense ab, we have the angle aRb : angle aCb :: CQ : RQ. Now as CQ is constant, and also the angle aRb being twice the refraction of the prism, the angle aCb varies as RQ. Hence the extent of the scale for measuring angles becomes the focal length of the object glass, and the angle measured is in proportion to the distance of the prisms from the principal focus of the object glass; and the micrometer can measure all angles (very small ones excepted, for the reason afterwards given) which do not exceed the sum of the refractions of the prisms; for the angle aCb, the diameter of the object to be measured, is always less than the angle aRb, the sum of the refractions of the prisms, except when the prisms touch the object glass, and then they become equal. The scale can never be out of adjustment, as the point a, where the measurement begins, answers to the focus of the object glass, which is a fixed point for all distant objects, and we have only to find the value of the scale answering to some known angle: for instance, bring the two limbs of the sun's images into contact, and measure the distance of the prisms from the focus, and look in the nautical almanac for the sun's diameter, and you get the value of the scale.

In fig. 8. the limb Q, of the image Q, is illuminated by the rays falling on the object glass between A and F, and of the image Q by those falling between B and G; but in fig. 9. the same limbs are illuminated by the rays falling between B and F, A and G respectively, and therefore will be more illuminated than in the other case; but the difference is not considerable on account of the great aperture of the object-glass compared with the distance FG.

It might be convenient to have two sets of prisms, one for measuring angles not exceeding 36', and therefore fit for measuring the diameters of the sun and moon, and the lucid parts and distances of the cusps in their eclipses; and another for measuring angles not much greater than 1', for the conveniency of measuring the diameters of the planets. For as QC : QR :: sum of the refractions of the prisms : angle aCb, the apparent diameter of the object, it is evident that if you diminish the third term, you must increase the second in the same ratio, in order to measure the same angle; and thus by diminishing the refractive angle of the prisms, you throw them further from Q, and consequently avoid the inconvenience of bringing them near to Q, for the reason in the next paragraph; and at the same time you will increase the illumination in a small degree. The prisms must be achromatic, each composed of two prisms of flint and crown glass, placed with their refracting angles contrariways, otherwise the images will be coloured.

In the construction here described, the angle measured becomes evanescent when the prisms come to the principal focus of the object glass, and therefore o on the scale then begins; but if the prisms be placed in the principal focus they can have no effect, because the pencil of rays at the junction of the prisms would then vanish, and therefore it is not practicable to bring the two images together to get o on the scale. Dr Maskelyne, therefore, thought of placing another pair of prisms within, to refract the rays before they came to the other prisms, by which means the two images would be formed into one before they came to the principal focus, and therefore o on the scale could be determined. But to avoid the error arising from the multiplication of mediums, he, instead of adding another pair of prisms, divided the object glass through its centre, and sliding the segments a little it separated the images, and then by the prisms he could form one image very distinctly, and consequently could determine o on the scale; for by separating the two segments you form two images, and you will separate the two pencils so that you may move up the two prisms, and the two pencils will fall on each respectively, and the two images may be formed into one. In the instrument which Dr Maskelyne had made, o on the scale was chosen to be about ¼ of the focal length of the object glass.

appears to coincide; but if we increase the refractive power of the eye, then each pencil is brought to a focus, and they cross each other before the rays come to the retina, consequently the two limbs on the retina will lap over; and if we diminish the refractive power of the eye, then each pencil being brought to a focus beyond the retina, and not crossing till after they have passed it, the two limbs on the retina must be separated.

glass, and each prism refracted 27°. By this means all angles are measured down to o.

In the Philosophical Transactions for 1779, Mr Ramsden has described two new micrometers, which he contrived with a view of remedying the defects of the object-glass micrometer.

1. One of these is a *catoptric* micrometer, which, beside the advantage it derives from the principle of reflection, of not being disturbed by the heterogeneity of light, avoids every defect of other micrometers, and can have no aberration, nor any defect arising from the imperfection of materials or of execution; as the extreme simplicity of its construction requires no additional mirrors or glasses to those required for the telescope; and the separation of the image being effected by the inclination of the two specula, and not depending on the focus of any lens or mirror, any alteration in the eye of an observer cannot affect the angle measured. It has peculiar to itself the advantages of an adjustment, to make the images coincide in a direction perpendicular to that of their motion; and also of measuring the diameter of a planet on both sides of the zero, which will appear no inconsiderable advantage to observers who know how much easier it is to ascertain the contact of the external edges of two images than their perfect coincidence.

A represents the small speculum divided into two equal parts; one of which is fixed on the end of the arm B; the other end of the arm is fixed on a steel axis X, which crosses the end of the telescope C. The other half of the mirror A is fixed on the arm D, which arm at the other end terminates in a socket *y*, that turns on the axis X; both arms are prevented from bending by the braces *a a*. G represents a double screw, having one part *e* cut into double the number of threads in an inch to that of the part *g*: the part *e* having 100 threads in one inch, and the part *g* 50 only. The screw *e* works in a nut F in the side of the telescope, while the part *g* turns in a nut H, which is attached to the arm B; the ends of the arms B and D, to which the mirrors are fixed, are separated from each other by the point of the double screw pressing against the stud *b*, fixed to the arm D, and turning in the nut H on the arm B. The two arms B and D are pressed against the direction of the double screw *eg* by a spiral spring within the part *n*, by which means all shake or play in the nut H, on which the measure depends, is entirely prevented.

From the difference of the threads on the screw at *e* and *g*, it is evident, that the progressive motion of the screw through the nut will be half the distance of the separation of the two halves of the mirror; and consequently the half mirrors will be moved equally in contrary directions from the axis of the telescope C.

The wheel V fixed on the end of the double screw has its circumference divided into 100 equal parts, and numbered at every fifth division with 5, 10, &c. to 100, and the index I shows the motion of the screw with the wheel round its axis, while the number of revolutions of the screw is shown by the divisions on the same index. The steel screw at R may be turned by the key S, and serves to incline the small mirror at right angles to the direction of its motion. By turning the finger-head T (fig. 11.), the eye-tube P is

brought nearer or farther from the small mirror, to adjust the telescope to distinct vision; and the telescope itself hath a motion round its axis for the conveniency of measuring the diameter of a planet in any direction. The inclination of the diameter measured with the horizon is shown in degrees and minutes by a level and vernier on a graduated circle, at the breech of the telescope.

" It is necessary to observe (says Mr Ramsden), that, besides the table for reducing the revolutions and parts of the screw to minutes, seconds, &c. it may require a table for correcting a very small error which arises from the excentric motion of the half-mirrors. By this motion their centres of curvature will (when the angle to be measured is large) approach a little towards the large mirror: the equation for this purpose in small angles is insensible; but when angles to be measured exceed ten minutes, it should not be neglected. Or, the angle measured may be corrected by diminishing it in the proportion the versed sine of the angle measured, supposing the eccentricity radius, bears to the focal length of the small mirror."

Mr Ramsden preferred Cassegrain's construction of the reflecting telescope to either the *Gregorian* or Newtonian; because in the former, errors caused by one speculum are diminished by those in the other. From a property of the reflecting telescope (which, he observes, has not been attended to), that the apertures of the two specula are to each other very nearly in the proportion of their focal lengths, it follows, that their aberrations will be to each other in the same proportion; and these aberrations are in the same direction, if the two specula are both concave; or in contrary directions, if one speculum is concave and the other convex. In the Gregorian construction, both specula being concave, the aberration at the second image will be the sum of the aberrations of the two mirrors; but in the Cassegrain construction, one mirror being concave and the other convex, the aberration at the second image will be the difference between their aberrations. By assuming such proportions for the foci of the specula as are generally used in the reflecting telescope, which is about as 1 to 4, the aberration in the Cassegrain construction will be to that in the Gregorian as 3 to 5.

2. The other is a *dioptric* micrometer, and suited to the principle of refraction. This micrometer is applied to the erect eye-tube of a refracting telescope, and is placed in the conjugate focus of the first eyeglass; in which position, the image being considerably magnified before it comes to the micrometer, any imperfection in its glass will be magnified only by the remaining eye-glasses, which in any telescope seldom exceeds five or six times. By this position also the size of the micrometer glass will not be the ⅟₂₀ part of the area which would be required if it was placed in the object-glass; and, notwithstanding this great disproportion of size, which is of great moment to the practical optician, the same extent of scale is preserved, and the images are uniformly bright in every part of the field of the telescope.

Fig. 11. represents the glasses of a refracting telescope *xy*, the principal pencil of rays from the object-glass O; *r r* and *v s*, the axis of two oblique pencils, *a*, the first eye-glass; *m*, its conjugate focus, or the place

Plate CCXCVI.

place of the micrometer; *b* the second eyeglass; *c* the third; and *d* the fourth, or that which is nearest the eye. Let *a* be the diameter of the object-glass, *c* the diameter of a pencil at *m*, and *f* the diameter of the pencil at the eye; it is evident, that the axis of the pencils from every part of the image will cross each other at the point *m* and *c*, the width of the micrometer-glass, is to *p* the diameter of the object-glass as *m* is to *p f*, which is the proportion of the magnifying power at the point *m*; and the error caused by an imperfection in the micrometer glass placed at *m* will be to the error, had the micrometer been at O, as *m* is to *p*.

Fig. 3. represents the micrometer: A, a convex or concave lens divided into two equal parts by a plane across its centre; one of these semi lenses is fixed in a frame B, and the other in the frame E; which two frames slide on a plate H, and are pressed against it by this plate: the frames B and E are moved in contrary directions by turning the button D; L is a scale of equal parts on the frame B; it is numbered from each end towards the middle with 10, 20, &c. There are two verniers on the frame E, one at M and the other at N, for the convenience of measuring the diameter of a planet, &c. on both sides the zero. The first division on both these verniers coincides at the same time with the two zeros on the scale L; and, if the frame is moved towards the right, the relative motion of the two frames is shown on the scale L by the vernier M; but if the frame B be moved towards the left, the relative motion is shown by the vernier N.—This micrometer has a motion round the axis of vision, for the convenience of measuring the diameter of a planet, &c. in any direction, by turning an endless screw F; and the inclination of the diameter measured with the horizon is shown on the circle *g* by a vernier on the plate V. The telescope may be adjusted to distinct vision by means of an adjusting screw, which moves the whole eye-tube with the micrometer nearer or farther from the object-glass, as telescopes are generally made; or the same effect may be produced in a better manner, without moving the micrometer, by sliding the part of the eye tube *a*, on the part *n*, by help of a screw or pinion. The micrometer is made to take off occasionally from the eye tube, that the telescope may be used without it.

Still, however, micrometers remained in several respects imperfect. In particular, the imperfections of the parallel-wire micrometer in taking the distance of very close double stars, are the following.

When two stars are taken between the parallels, the diameters will be included. Mr Herschel informs us, he has in vain attempted to find lines sufficiently thin to extend them across the centres of the stars so that their thickness might be neglected. The single threads of the silk-worm, with such lenses as he uses, are so much magnified that their diameter is more than that of many of the stars. Besides, if they were much less than they are, the power of deflection of light would make the attempt to measure the distance of the centres this way fruitless; for he has always found the light of the stars to play upon those lines and separate their apparent diameters into two parts. Now since the spurious diameters of the stars thus included, as Mr Herschel assures us, are continually changing according to the state of the air,

and the length of time we look at them, we are, in some respect, left at an uncertainty, and our measures taken at different times and with different degrees of attention, will vary on that account. Nor can we come at the true distance of the centres of any two stars, one from another, unless we could tell what to allow for the semidiameters of the stars themselves; for different stars have different apparent diameters, which, with a power of 227, may differ from each other as far as two seconds.

The next imperfection is that which arises from a deflection of light upon the wires when they approach very near to each other; for if this be owing to a power of repulsion lodged at the surface, it is easy to understand, that such powers most interfere with each other, and give the measures larger in proportion than they would have been if the repulsive power of one wire had not been opposed by a contrary power of the other wire.

Another very considerable imperfection of these micrometers is a continual uncertainty of the real zero. Mr Herschel has found, that the least alteration in the situation and quantity of light will affect the zero, and that a change in the position of the wires, when the light and other circumstances remain unaltered, will also produce a difference. To obviate this difficulty, whenever he took a measure that required the utmost accuracy, his zero was always taken immediately after, while the apparatus remained in the same situation it was in when the measure was taken; but this enhances the difficulty, because it introduces an additional observation.

The next imperfection, which is none of the smallest, is that every micrometer that has hitherto been in use requires either a screw, or a divided bar and pinion, to measure the distance of the wires or divided image. Those who are acquainted with works of this kind are but too sensible how difficult it is to have screws that shall be perfectly equal in every thread or revolution of each thread; or pinions and bars that shall be so evenly divided as perfectly to be depended upon in every leaf and tooth to perhaps the two, three, or four thousandth part of an inch; and yet, on account of the small scale of those micrometers, these quantities are of the greatest consequence; an error of a single thousandth part inducing in most instruments a mistake of several seconds.

The last and greatest imperfection of all is, that these wire micrometers require a pretty strong light in the field of view; and when Mr Herschel had double stars to measure, one of which was very obscure, he was obliged to be content with less light than is necessary to make the wires perfectly distinct; and several stars on this account could not be measured at all, though otherwise not too close for the micrometer.

Mr Herschel, therefore, having long had much occasion for micrometers that would measure exceeding small distances exactly, was led to bend his attention to the improvement of these instruments; and the result of his endeavours has been a very ingenious instrument called a *lamp-micrometer*, which is not only free from the imperfections above specified, but also possesses the advantages of a very large scale. This instrument is described in the Philosophical Transactions for 1782; and the construction of it is as follows:

ABGCFE

Plate CCXCVI.

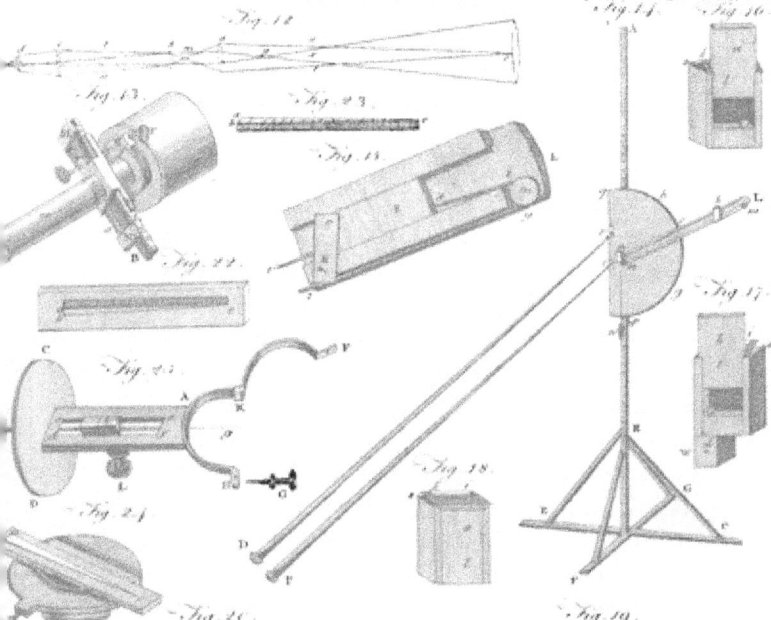

Fig. 12

Fig. 13

Fig. 23

Fig. 15

Fig. 22

Fig. 14

Fig. 16

Fig. 17

Fig. 25

Fig. 24

Fig. 26

Fig. 18

Fig. 19

Fig. 20

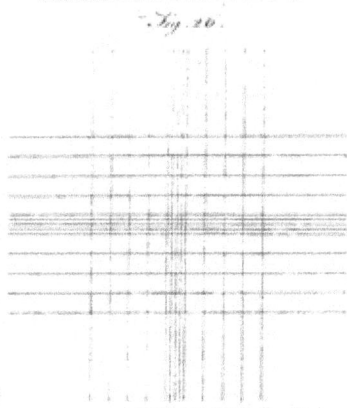

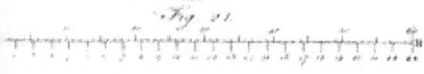

Fig. 21

J. Bell Prin Med Sculptor fecit.

Micrometer.

ABGCFE (fig. 14.) is a stand nine feet high, upon which a semicircular board *qbcpg* is moveable upwards or downwards, in the manner of some fire-screens, as occasion may require, and is held in its situation by a peg *p* put into any one of the holes of the upright piece *B*. This board is a segment of a circle of fourteen inches radius, and is about three inches broader than a semicircle, to give room for the handles *D*, *P*, to work. The use of this board is to carry an arm L, thirty inches long, which is made to move upon a pivot at the centre of the circle, by means of a string, which passes in a groove upon the edge of the semicircle *pqabq*; the string is fastened to a hook at *o* (not expressed in the figure, being at the back of the arm L), and passing along the groove from *ab* to *q* is turned over a pulley at *q*, and goes down to a small barrel *r*, within the plane of the circular board, where a double-jointed handle *P* commands its motion. By this contrivance, we see, the arm L may be lifted up to any altitude from the horizontal position to the perpendicular, or be suffered to descend by its own weight below the horizontal to the reverse perpendicular situation. The weight of the handle P is sufficient to keep the arm in any given position; but if the motion should be too easy, a friction spring applied to the barrel will moderate it at pleasure.

In front of the arm L, a small slider, about three inches long, is moveable in a rabbet from the end L towards the centre backwards and forwards. A string is fastened to the left side of the little slider, and goes towards L, where it passes round a pulley at *m*, and returns under the arm from *m*, *n*, towards the centre, where it is led in a groove on the edge of the arm, which is of a circular form, upwards to a barrel (raised above the plane of the circular board) at *r*, to which the handle *D* is fastened. A second string is fastened to the slider, at the right side, and goes towards the centre, where it passes over a pulley *n*; and the weight *w*, which is suspended by the end of this string, returns the slider towards the centre, when a contrary turn of the handle permits it to act.

By *a* and *b* are represented two small lamps, two inches high, 1½ in breadth by 1½ in depth. The sides, back, and top, are made so as to permit no light to be seen, and the front consists of a thin brass sliding door. The flame in the lamp *a* is placed three-tenths of an inch from the left side, three-tenths from the front, and half an inch from the bottom. In the lamp *b* it is placed at the same height and distance, measuring from the right side. The wick of the flame consists only of a single very thin lamp cotton-thread; for the smallest flame being sufficient, it is easier to keep it burning in so confined a place. In the top of each lamp must be a little slit lengthways, and also a small opening in one side near the upper part, to permit air enough to circulate to feed the flame. To prevent every reflection of light, the side opening of the lamp *a* should be to the right, and that of the lamp *b* to the left. In the sliding door of each lamp is made a small hole with the point of a very fine needle just opposite the place where the wicks are burning, so that when the sliders are shut down, and every thing dark, nothing shall be seen but two fine lucid points of the fire of two stars of the third or fourth magnitude. The lamp *a* is placed so that its lucid point may be in the centre of

the circular board where it remains fixed. The lamp *b* is hung to the little slider which moves in the rabbet of the arm, so that its lucid point, in an horizontal position of the arm, may be on a level with the lucid point in the centre. The moveable lamp is suspended upon a piece of brass fastened to the slider by a pin exactly behind the flame, upon which it moves as a pivot. The lamp is balanced at the bottom by a leaden weight, so as always to remain upright, when the arm is either lifted above or depressed below the horizontal position. The double-jointed handles *D*, *P*, consist of light deal rods, ten feet long, and the lowest of them may have divisions, marked upon it near the end P, expressing exactly the distance from the central lucid point in feet, inches, and tenths.

From this construction we see, that a person at a distance of ten feet may govern the two lucid points, so as to bring them into any required position south or north preceding or following from o to 90° by using the handle P, and also to any distance from six-tenths of an inch to five or six and twenty inches by means of the handle D. If any reflection or appearance of light should be left from the top or sides of the lamps, a temporary screen, consisting of a long piece of pasteboard, or a wire frame covered with black cloth, of the length of the whole arm, and of any required breadth, with a slit of half an inch broad in the middle, may be affixed to the arm by four bent wires projecting an inch or two before the lamps, situated to that the moveable lucid point may pass along the opening left for that purpose.

Fig. 15. represents part of the arm L, half the real size; *b* the slider; *n* the pulley, over which the cord *abyr* is returned towards the centre; *v* the other cord going to the pulley *n* of fig. 14. R the brass piece moveable upon the pin *r*, to keep the lamp upright. At R is a wire fretted to the brass piece, upon which is held the lamp by a nut and screw. Fig 16. 17. represent the lamps *a*, *b*, with the sliding doors open, to show the situation of the wicks. W is the leaden weight with a hole of its in, through which the wire R of fig. 15. is to be pushed when the lamp is to be fast, and to the slider S. Fig. 18. represents the lamp *a* with the sliding door shut; *t* the lucid point; and *s* the openings at the top, and *s* at the sides, for the admission of air.

" Every ingenious artist (says Mr Herschel) will soon perceive, that the motions of this micrometer are capable of great improvement by the application of wheels and pinions, and other well known mechanical resources; but as the principal object is only to be able to adjust the two lucid points to the required position and distance, and to keep them there for a few minutes, while the observer goes to measure their distance, it will not be necessary to say more upon the subject.

" I am now to show the application of this instrument. It is well known to opticians and others who have been in the habit of using optical instruments, that we can with one eye look into a microscope or telescope, and see an object much magnified, while the naked eye may see a scale upon which the magnified picture is thrown. In this manner I have generally determined the power of my telescopes; and any one who has acquired a facility of taking such observations

will

Micrometer.

will very seldom mistake so much as one in fifty in determining the power of an instrument, and that degree of exactness is fully sufficient for the purpose.

"The Newtonian form is admirably adapted to the use of this micrometer; for the observer stands always erect, and looks in a horizontal direction, notwithstanding the telescope should be elevated to the zenith. Besides, his face being turned away from the object to which his telescope is directed, this micrometer may be placed very conveniently without causing the least obstruction to the view: therefore, when I use this instrument, I put it at ten feet distance from the left eye, in a line perpendicular to the tube of the telescope, and raise the moveable board to such a height that the lucid point of the central lamp may be upon a level with the eye. The handles, lifted up, are passed through two loops fastened to the tube, just by the observer, so as to be ready for his use. I should observe, that the end of the tube is cut away, so as to leave the left eye entirely free to see the whole micrometer.

"Having now directed the telescope to a double star, I view it with the right eye, and at the same time with the left see it projected upon the micrometer: then, by the handle P, which commands the position of the arm, I raise or depress it so as to bring the two lucid points to a similar situation with the two stars; and, by the handle D, I approach or remove the moveable lucid point to the same distance of the two stars, so that the two lucid points may be exactly covered by or coincide with the stars. A little practice in this business soon makes it easy, especially to one who has already been used to look with both eyes open.

"What remains to be done is very simple. With a proper rule, divided into inches and fortieth parts, I take the distance of the lucid points, which may be done to the greatest nicety, because, as I observed before, the little holes are made with the point of a very fine needle. The measure thus obtained is the tangent of the magnified angle under which the stars are seen to a radius of ten feet; therefore, the angle being found and divided by the power of the telescope gives the real angular distance of the centres of a double star.

"For instance, September 23, 1781, I measured α Herculis with this instrument. Having caused the two lucid points to coincide exactly with the stars centre upon centre, I found the radius or distance of the central lamp from the eye 10 feet 4.15 inches; the tangent or distance of the two lucid points 50.0 fortieth parts of an inch; this gives the magnified angle 35′, and dividing by the power 460, which I used, we obtain 4″ 34‴ for the distance of the centres of the two stars. The scale of the micrometer at this very convenient distance, with the power of 460 (which my telescope bears so well upon the fixed stars that for near a twelvemonth past I have hardly used any other), is above a quarter of an inch to a second; and by putting on my power of 932, which in very fine evenings is extremely distinct, I obtain a scale of more than half an inch to a second, without increasing the distance of the micrometer; whereas the most perfect of my former micrometers, with the same instrument, had a scale of less than the two thousandth part of an inch to a second.

"The measures of this micrometer are not confined to double stars only, but may be applied to any other objects that require the utmost accuracy, such as the diameters of the planets or their satellites, the mountains of the moon, the diameters of the fixed stars, &c.

"For instance, October 22, 1781, I measured the apparent diameter of α Lyræ; and judging it of the greatest importance to increase my scale as much as convenient, I placed the micrometer at the greatest convenient distance, and (with some trouble, for want of longer handles, which might easily be added) took the diameter of this star by removing the two lucid points to such a distance as just to exclude the apparent diameter. When I measured my radius, it was found to be twenty-two feet six inches. The distance of the two lucid points was about three inches, for I will not pretend to extreme nicety in this observation, on account of the very great power I used, which was 6450. From these measures we have the magnified angle 38′ 10″; this divided by the power gives 0 .355 for the apparent diameter of α Lyræ. The scale of the micrometer, on this occasion, was no less than 8.443 inches to a second, as will be found by multiplying the natural tangent of a second with the power and radius in inches.

"November 28, 1781, I measured the diameter of the new star; but the air was not very favourable, for this singular star was not so distinct with 227 that evening as it generally is with 460: therefore, without laying much stress upon the exactness of the observation, I shall only repeat it to exemplify the use of the micrometer. My radius was 53 feet 11 inches. The diameter of the star, by the distance of the lucid points, was 2.4 inches, and the power I used 227: hence the magnified angle is found 14′, and the real diameter of the star 0″.022. The scale of this measure .474 millesimals of an inch, or almost half an inch to a second."

In the Philosophical Transactions for 1791, a very simple micrometer for measuring small angles with the telescope is described by Mr Cavallo; who introduces his description with the following observations upon the different sorts of telescopical micrometers in use. "These instruments may be divided into two classes; namely, those which have not, and those which have, some movement amongst their parts. The micrometers of the former sort consist mostly of fine wires or hairs, variously disposed, and situated within the telescope, just where the image of the object is formed. In order to determine an angle with those micrometers, a good deal of calculation is generally required. The micrometers of the other sort, of which there is a great variety, some being made with moveable parallel wires, others with prisms, others again with a combination of lenses, and so on, are more or less subject to several inconveniencies, the principal of which are the following. 1. Their motions generally depend upon the action of a screw; and of course the imperfections of its threads, and the greater or less quantity of lost motion, which is observable in moving a screw, especially when small, occasion a considerable error in the mensuration of angles. 2. Their complication and bulk renders them difficultly applicable to a variety of telescopes, especially to the pocket ones. 3. They do not measure the angle without some loss of time, which is necessary to turn the screw,

Microme-
ter.

forew, or to move some other mechanism. 4. and
lastly, They are considerably expensive, so that some
of them cost even more than a tolerably good telescope.

After having had long in view (one author informs
us) the construction of a micrometer which might
be in part at least, if not entirely, free from all those
objections; he, after various attempts, at last succeeded with a simple contrivance, which, after repeated trials, has been found to answer the desired
end, not only from his own experience, but from that
also of several friends, to whom it has been communi-
cated.

This micrometer, in short, consists of a thin and
narrow slip of mother-of-pearl finely divided, and situated in the focus of the eye-glass of a telescope, just
where the image of the object is formed. It is imma-
terial whether the telescope be a refractor or a reflector, provided the eye-glass be a convex lens, and not
a concave one as in the Galilean construction.

The simplest way of fixing it is to stick it upon
the diaphragm which generally stands within the
tube and in the focus of the eye-glass. When then
in fact, if you look through the eye-glass, the divisions
of the micrometrical scale will appear very distinct,
unless the diaphragm is not exactly in the focus; in
which case, the micrometrical scale must be placed
exactly in the focus of the eye-glass, either by pushing the diaphragm backwards or forwards, when that
is practicable; or else the scale may be easily removed
from one or the other surface of the diaphragm by
the interposition of a circular piece of paper or card,
or by a bit of wax. This construction is fully sufficient, when the telescope is always to be used by the
same person; but when different parties are to use it,
then the diaphragm which supports the micrometer
must be constructed so as to be easily moved backwards
or forwards, though that motion needs not be greater
than about a tenth or an eighth of an inch. This is
necessary, because the distance of the focus of the
same lens appears different to the eyes of different
persons; and, therefore, whoever is going to use the
telescope for the measuration of any angle, must first
of all unscrew the tube which contains the eye-glass
and micrometer from the rest of the telescope, and,
looking through the eye-glass, must place the micrometer where the divisions of it may appear quite distinct to his eye.

In case that any person should not like to see always the micrometer in the field of the telescope, then
the micrometrical scale, instead of being fixed to the
diaphragm, may be fitted to a circular perforated
plate of brass, wood, or even paper, which may be,
occasionally placed upon the said diaphragms.

Mr Cavallo has made several experiments to determine the most useful substance for this micrometer.—
Glass, which he had successfully applied for a similar
purpose to the tangental microscope, seemed at first
to be the most promising; but it was at last rejected
after several trials; for the divisions upon it generally
are either too fine to be perceived, or too rough; and
though with proper care and attention the divisions
may be proportioned to the sight, yet the thickness
of the glass itself obstructs in some measure the distinct

view of the object. Ivory, horn, and wood, were
found useless for the construction of this micrometer,
on account of their bending, swelling, and contracting very easily; whereas mother-of-pearl is a very
steady substance, the divisions upon it may be marked
very easily, and when it is made as thin as common
writing paper it has a very useful degree of transparency.

Fig. 19. exhibits this micrometer scale, but shows
it four times larger than the real size of one, which
he has adapted to a three-feet achromatic telescope
that magnifies about 84 times. It is something less than
the 24th part of an inch broad; its thickness is equal
to that of common writing-paper; and the length of
it is determined by the aperture of the diaphragm,
which limits the field of the telescope. The divisions
upon it are the 20ths of an inch, which reach from
one edge of the scale to about the middle of it, excepting every fifth and tenth division, which are
longer. The divided edge of it passes through the
centre of the field of view, though this is not a necessary precaution in the construction of this micrometer.
Two divisions of the above described scale in my telescope are very nearly equal to one minute; and as a
quarter of one of those divisions may be very well distinguished by estimation, therefore an angle of one
eighth part of a minute, or of 7½", may be measured
with it.

When a telescope magnifies more, the divisions of
the micrometer must be more minute; and Mr Cavallo finds, that when the focus of the eye-glass of the
telescope is shorter than half an inch, the micrometer
may be divided with the 500ths of an inch; by
means of which, and the telescope magnifying about
200 times, one may easily and accurately measure an
angle smaller than half a second. On the other hand,
when the telescope does not magnify above 20 times,
the divisions used may be so minute; for instance, in
one of Dollond's pocket telescopes, which when drawn
out for use is about 14 inches long, a micrometer with
the hundredths of an inch is quite sufficient, and one
of its divisions is equal to little less than three minutes, so that an angle of a minute may be measured
by it.

" In looking through a telescope furnished with
such a micrometer (says our author), the field of view
appears divided by the micrometer scale, the breadth
of which occupies about one-seventh part of the aperture; and as the scale is semitransparent, that part of
the object which happens to be behind it may be discerned (sufficiently well to ascertain the division, and
even the quarter of a division, with which its borders
coincide. Fig. 20. shows the appearance of the field
of my telescope with the micrometer, when directed
to the title page of the Philosophical Transactions,
wherein one may observe that the thickness of the
letter C is equal to three-fourths of a division, the
diameter of the O is equal to three divisions, and
so on.

" At first view, one is apt to imagine, that it is difficult to count the divisions which may happen to cover or to measure an object; but upon trial it will be
found, that this is readily performed; and even people
who have never been used to observe with the tele-

Microme-
ter.

Microme-
ter.

scope, foon learn to meafure very quickly and accurately with this micrometer; for fince every fifth and
tenth divifion is longer than the reft, one foon acquires
the habit of faying, five, ten, fifteen; and then, by
adding the other divifions lefs than five, completes
the reckoning. Even with a telefcope which has no
ftand, if the object end of it be refted againft a fteady
place, and the other end be held by the hand near
the eye of the obferver, an object may be meafured
with accuracy fufficient for feveral purpofes, as for the
estimation of fmall diftances, for determining the
height of a houfe, &c.

"After having conftructed and adapted this micrometer to the telefcope, it is then neceffary to afcertain the value of the divifions. It is hardly neceffary
to mention in this place, that though thefe divifions
meafure the chords of the angles, and not the angles
or arches themfelves, and the chords are not as the
arches, yet it has been fhown by all the trigonome-
trical writers, that in fmall angles the chords, arches,
fines, and tangents, follow the fame proportion fo
very nearly, that the very minute difference may be
fafely neglected; fo that if one divifion of this micrometer is equal to one minute, we may fafely conclude,
that two divifions are equal to two minutes, three divifions to three minutes, and fo on. There are various methods of afcertaining the value of the divifions
of fuch a micrometer, they being the very fame that
are ufed for afcertaining the value of the divifions in
other micrometers. Such are, the paffage of an equatorial ftar over a certain number of divifions in a certain time; or the meafuring of the diameter of the fun,
by computation from the focal diftance of the object
and other lenfes of the telefcope; the laft of which,
however, is fubject to feveral inaccuracies; but as
they are well known to aftronomical perfons, and
have been defcribed in many books, they need not
be farther noticed here. However, for the fake of
workmen and other perfons not converfant in aftro-
nomy, I fhall defcribe an eafy and accurate method of
afcertaining the value of the divifions of the micrometer.

"Mark upon a wall or other place the length of
fix inches, which may be done by making two dots
or lines fix inches afunder, or by fixing a fix-inch ruler upon a ftand; then place the telefcope before it
fo that the ruler or fix-inch length may be at right
angles with the direction of the telefcope, and juft 57
feet 3½ inches diftant from the object glafs of the
telefcope: this done, look through the telefcope at
the ruler or other extenfion of fix inches, and obferve
how many divifions of the micrometer are equal to
it, and that fame number of divifions is equal to half
a degree, or 30'; and this is all that needs be done
for the required determination; the reafon of which
is, becaufe an extenfion of fix inches fubtends an
angle of 30' at the diftance of 57 feet 3½ inches, as
may be eafily calculated by the rules of plane trigo-
nometry.

"In one of Dollond's 14-inch pocket telefcopes,
if the divifions of the micrometer be the hundredths of
an inch, 11½ of thefe divifions will be found equal to
30', or 23 to a degree. When this value has been
once afcertained, any other angle meafured by any
other number of divifions is determined by the rule

of three. Thus, fuppofe that the diameter of the fun,
feen through the fame telefcope, be found equal to
12 divifions, fay as 11½ divifions are to 30 minutes,

fo are 12 divifions to $\left(\frac{12 \times 30}{11\frac{1}{2}}\right)$ 31'·3, which is the

required diameter of the fun.

"Notwithftanding the facility of this calculation, a
fcale may be made anfwering to the divifions of a
micrometer, which will fhew the angle correfponding
to any number of divifions to mere infpection. Thus,
for the above-mentioned fmall telefcope, the fcale is
reprefented in fig. 21. AB is a line drawn at pleafure; it is then divided into 23 equal parts, and thofe
divifions which reprefent the divifions of the micrometer that are equal to one degree, are marked on
one fide of it. The line then is divided again into
60 equal parts, which are marked on the other fide
of it; and thefe divifions reprefent the minutes which
correfpond to the divifions of the micrometer; thus
the figure fhews, that fix divifions of the micrometer
are equal to 15½ minutes, 11½ divifions are nearly
equal to 29 minutes, &c. What has been faid of minutes may be faid of feconds alfo, when the fcale is
to be applied to a large telefcope.

"Thus far this micrometer and its general ufe have
been fufficiently defcribed; and mathematical perfons
may eafily apply it to the various purpofes to which
micrometers have been found fubfervient. But as the
fimplicity, cheapnefs, and at the fame time the accuracy of this contrivance, may render the ufe of it much
more general than that of any other micrometer; and
I may venture to fay, that it will be found very ufeful in the army, and amongft fea-faring people, for
the determination of diftances, heights, &c. I fhall
therefore join fome practical rules to render this micrometer ufeful to perfons unacquainted with trigo-
nometry and the ufe of logarithms.

"Problem I. The angle, not exceeding one degree,
which is fubtended by an extenfion of our foot, being
given, to find its diftance from the place of obfervation. N. B. This extenfion of one foot, or any other
which may be mentioned hereafter, muft be perpendicular to the direction of the telefcope through which
it is obferved. The diftances are reckoned from the
object-glafs of the telefcope; and the anfwers obtained
by the rules of this problem, though not exactly true,
are however fo little different from the truth, that the
difference feldom amounts to more than two or three
inches, which may be fafely neglected.

"Rule 1. If the angle be expreffed in minutes,
fay, as the given angle is to 60, fo is 1875.55 to a
fourth proportional, which gives the anfwer in inches.
—2. If the angle be expreffed in feconds, fay, as
the given angle is to 3600, fo is 687.55 to a fourth
proportional, which expreffes the anfwer in inches.
—3. If the angle be expreffed in minutes and feconds, turn it all into feconds, and proceed as a-
bove.

"Example. At what diftance is a globe of one
foot in diameter when it fubtends an angle of two
feconds?

2 : 3600 :: 687.55 : $\frac{3600 \times 687.55}{2}$ = 1237590

inches, or 103132½ feet, which is the anfwer required.

This

Microme-
ter.

" This calculation may be shortened; for since two of the three proportionals are fixed, their product in the first case is 41253, and in the other two cases is .2475180; so that in the first case, viz. when the angle is expressed in minutes, you need only divide 41253 by the given angle; and in the other two cases, viz. when the angle is expressed in seconds, divide 2475180 by the given angle, and the quotient in either case is the answer in inches.

" Problem II. The angle, not exceeding one degree, which is subtended by any known extension, being given, to find its distance from the place of observation.

" Rule. Proceed as if the extension were of one foot by Problem I. and call the answer B; then, if the extension in question be expressed in inches, say, as 12 inches are to that extension, so is B to a fourth proportional, which is the answer in inches; but if the extension in question be expressed in feet, then you need only multiply it by B, and the product is the answer in inches.

" Example. At what distance is a man six feet high, when he appears to subtend an angle of 30'.

" By problem I. if the man were one foot high, the distance would be 82506 inches; but as he is six feet high, therefore multiply 82506 by 6, and the product gives the required distance, which is 495036 inches, or 41253 feet.

" For greater conveniency, especially in travelling, or in such circumstances in which one has not the opportunity of making even the easy calculations required in these problems, I have calculated the following two tables; the first of which shows the distance answering to any angle from one minute to one degree, which is subtended by an extension of one foot; and the second table shows the distance answering to any angle from one minute to one degree, which is subtended by a man, the height of which has been called an extension of six feet; because, at a mean, such is the height of a man when dressed with hat and shoes on. These tables may be transferred on a card, and may be had always ready with a pocket telescope furnished with a micrometer. Their use is evidently to ascertain distances without any calculation; and they are calculated only to minutes, because with a pocket telescope and micrometer it is not possible to measure an angle more accurately than to a minute.

" Thus, if one wants to measure the extension of a street, let a foot ruler be placed at the end of the street; measure the angular appearance of it, which suppose to be 30', and in the table you will have the required distance against 30', which is 95½ feet. Thus also a man who appears to be 40' high, is at the distance of 421 feet.

Angles subtended by an extension of one foot at different distances.

Angles	Distances in feet.	Angles	Distances in feet.
Min. 1	3437,7	Min. 31	110,9
2	1718,9	32	107,4
3	1145,9	33	104,2
4	859,4	34	101,1
5	687,5	35	98,2
6	572,9	36	95,5
7	491,1	37	92,9
8	429,7	38	90,4
9	382,0	39	88,1
10	343,7	40	85,9
11	312,5	41	83,8
12	286,5	42	81,8
13	264,4	43	79,9
14	245,5	44	78,1
15	229,2	45	76,4
16	214,8	46	74,7
17	202,2	47	73,1
18	190,9	48	71,6
19	180,9	49	70,1
20	171,8	50	68,7
21	163,7	51	67,4
22	156,2	52	66,1
23	149,4	53	64,8
24	143,2	54	63,6
25	137,5	55	62,5
26	132,2	56	61,4
27	127,3	57	60,3
28	122,7	58	59,2
29	118,5	59	58,2
30	114,6	60	57,3

Angles subtended by an extension of six feet at different distances.

Angles	Distances in feet.	Angles	Distances in feet.
Min. 1	20626,5	Min. 31	665,4
2	10313,	32	644,5
3	6875,4	33	625,
4	5156,6	34	606,6
5	4125,3	35	589,3
6	3437,7	36	572,9
7	2946,6	37	557,5
8	2578,3	38	542,8
9	2291,8	39	528,9
10	2062,6	40	515,6
11	1875,1	41	503,1
12	1718,8	42	491,1
13	1586,7	43	479,7
14	1473,3	44	468,8
15	1375,	45	458,4
16	1289,2	46	448,4
17	1213,3	47	438,9
18	1145,9	48	429,7
19	1085,6	49	421,
20	1031,4	50	412,5
21	982,2	51	404,4
22	937,6	52	396,7
23	896,8	53	389,2
24	859,4	54	381,9
25	825,	55	375,
26	793,3	56	368,3
27	763,9	57	361,9
28	736,6	58	355,6
29	711,3	59	349,6
30	687,5	60	343,7

II. The Micrometer has not only been applied to telescopes, and employed for astronomical purposes; but
there have also been various contrivances for adapting
it to MICROSCOPICAL observations. Mr Leeuwen-
hoek's method of estimating the size of small objects
was by comparing them with grains of sand, of which
100 in a line took up an inch. These grains he laid
upon the same plate with his objects, and viewed them
at the same time. Dr Jurin's method was similar to
this; for he found the diameter of a piece of fine filver wire, by wrapping it as close as he could about a
pin, and observing how many rings made an inch;
and he used this wire in the same manner as Leeuwen-
hoek used his sand. Dr Hooke used to look upon the
magnified object with one eye, while at the same time
he viewed other objects placed at the same distance
with the other eye. In this manner he was able, by
the help of a ruler, divided into inches and small parts,
and laid on the pedestal of the microscope, to cast as
it were the magnified appearance of the object upon
the ruler, and thus exactly to measure the diameter
which it appeared to have through the glass; which
being compared with the diameter as it appeared to
the naked eye, easily showed the degree in which it
was magnified. A little practice, says Mr Baker,
will render this method exceedingly easy and pleasant.

Mr Martin in his Optics recommended such a micrometer for a microscope as had been applied to telescopes; for he advises to draw a number of parallel
lines on a piece of glass, with the fine point of a diamond, at the distance of one-fortieth of an inch from
one another, and to place it in the focus of the eye-
glass. By this method, Dr Smith contrived to take
the exact draught of objects viewed by a double microscope; for he advises to get a lattice, made with
small silver wires or squares, drawn upon a plain glass
by the strokes of a diamond, and to put it into the
place of the image, formed by the object-glass: then
by transferring the parts of the object, seen in the
squares of the glass or lattice upon similar corresponding squares drawn on paper, the picture may be exactly taken. Mr Martin also introduced into compound
microscopes another micrometer, consisting of a screw.
See both these methods described in his Optics, p. 277.

The mode of actual admeasurement (Mr Adams
observes*) is without doubt the most simple that can
be used; as by it we comprehend, in a manner, at
one glance, the different effects of combined glasses;
and as it saves the trouble, and avoids the obscurity,
of the usual modes of calculation: but many persons
find it exceedingly difficult to adopt this method, because they have not been accustomed to observe with
both eyes at once. To obviate this inconvenience,
the late Mr Adams contrived an instrument called the
Needle-Micrometer, which was first described in his Mi-
crographia Illustrata; and of which, as now constructed,
we have the following description by his son Mr George
Adams in the ingenious Essays above quoted.

This micrometer consists of a screw, which has 50
threads to an inch; this screw carries an index, which
points to the divisions on a circular plate, which is
fixed at right angles to the axis of the screw. The
revolutions of the screw are counted on a scale, which
is an inch divided into 50 parts; the index to these divisions is a flower-de-luce marked upon the slider, which

* Microgra-
phical Essays,
p. 19.

carries the needle point across the field of the microscope. Every revolution of the micrometer screw
measures $\frac{1}{50}$th part of an inch, which is again subdi-
vided by means of the divisions on the circular plate,
as this is divided into 20 equal parts, over which the
index passes at every revolution of the screw; by which
means we obtain with ease the measure of $\frac{1}{1000}$th part
of an inch: for 50, the number of threads on the
screw in one inch, being multiplied by 20, the divisions on the circular plate are equal to 1000; so that
each division on the circular plate shows that the
needle has either advanced or receded 1000th part of
an inch.

To place this micrometer on the body of the microscope, open the circular part FKH, fig. 25. by taking
out the screw G, throw back the semicircle FK, which
moves upon a joint at K; then turn the sliding tube of
the body of the microscope, so that the small holes which
are in both tubes may exactly coincide, and let the needle
g of the micrometer have a free passage through them;
after this, screw it fast upon the body by the screw G.
The needle will now traverse the field of the microscope, and measure the length and breadth of the
image of any object that is applied to it. But further assistance must be had, in order to measure the
object itself, which is a subject of real importance;
for though we have ascertained the power of the microscope, and know that it is so many thousand times,
yet this will be of little assistance towards ascertaining
an accurate idea of its real size; for our ideas of bulk
being formed by the comparison of one object with
another, we can only judge of that of any particular
body, by comparing it with another whose size is
known: the same thing is necessary, in order to form
an estimate by the microscope; therefore, to ascertain
the real measure of the object, we must make the point
of the needle pass over the image of a known part of
an inch placed on the stage, and write down the revolutions made by the screw, while the needle passed
over the image of this known measure; by which
means we ascertain the number of revolutions on the
screw, which are adequate to a real and known measure on the stage. As it requires an attentive eye to
watch the motion of the needle point as it passes over
the image of a known part of an inch on the stage, we
ought not to trust to one single measurement of the
image, but ought to repeat it at least six times; then
add the six measures thus obtained together, and divide their sum by six, or the number of trials; the
quotient will be the mean of all the trials. This refult
is to be placed in a column of a table next to that
which contains the number of the magnifiers.

By the assistance of the sectoral scale, we obtain
with ease a small part of an inch. This scale is shown
at fig. 22, 23, 24, in which the two lines ca, cb, with
the side ab, form an isosceles triangle; each of the
sides is two inches long, and the base still only of
one-tenth of an inch. The longer sides may be of any
given length, and the base still only of one-tenth of an
inch. The longer lines may be considered as the line
of lines upon a sector opened to one-tenth of an inch.
Hence whatever number of equal parts ca, cb are divided into, their transverse measure will be such a part
of one-tenth as is expressed by their divisions. Thus
if it be divided into two equal parts, this will divide
the

Microme-
ter.

the inch into 100 equal parts; the first division next
c will be equal to 100th part of an inch, because it is
the tenth part of one-tenth of an inch. If these lines
are divided into twenty equal parts, the inch will be
by that means divided into 200 equal parts. Lastly,
if ab, ca, are made three inches long, and divided into
100 equal parts, we obtain with ease the 1000th part.
The scale is represented as folid at fig. 23, but as perforated at fig. 22, and 24, fo that the light paffes thro'
the aperture, when the fectoral part is placed on the ftage.

To ufe this fcale, firft fit the micrometer, fig. 25,
to the body of the microfcope; then fit the fectoral
fcale, fig. 24, in the ftage, and adjuft the microfcope
to its proper focus or diftance from the fcale, which is
to be moved till the bafe appears in the middle of the
field of view; then bring the needle point g, fig. 25,
(by turning the fcrew I,) to touch one of the lines ca,
exactly at the point anfwering 10 20 on the fectoral
fcale. The index a of the micrometer is to be fet to
the firft divifion, and that on the dial-plate to 20, which
is both the beginning and end of its divifions; we are
then prepared to find the magnifying power of every
magnifier in the compound microfcope which we are
ufing.

Example. Every thing being prepared agreeable to
the foregoing directions, fuppofe you are defirous of
afcertaining the magnifying power of the lens marked
N° 4. turn the micrometer fcrew until the point of
the needle has paffed over the magnified image of the
tenth part of one inch; then the divifion, where the
two indices remain, will fhow how many revolutions,
and parts of a revolution, the fcrew has made, while
the needle point traverfed the magnified image of the
one-tenth of an inch; fuppofe the refult to be 26 revolutions of the fcrew, and 14 parts of another revolution, this is equal to 26 multiplied by 20, added to 14;
that is, 134,000 parts of an inch.—The 26 divifions
found on the ftraight fcale of the micrometer, while
the point of the needle paffed over the magnified
image of one-tenth part of an inch, were multiplied
by 20, becaufe the circular plate CD, fig. 25, is divided into 20 equal parts; then produced 520; then
adding the 14 parts of the next revolution, we obtain
the 534,000 parts of an inch, or five-tenths and 3400
parts of another tenth, which is the meafure of the
magnified image of one-tenth of an inch, at the aperture of the eye-glaffes as at their fuci. Now if we
fuppofe the focus of the two eye-glaffes to be one
inch, the double thereof is two inches; or if we reckon in the 1000th part of an inch, we have 2000 parts
for the diftance of the eye from the needle point of
the micrometer. Again, if we take the diftance of
the image from the object at the ftage at 6 inches, or
6000, and add thereto 2000, double the diftance of
the focus of the eye-glafs, we fhall have 8000 parts of
an inch for the diftance of the eye from the object; and
as the glaffes double the image, we muft double the number 534 found upon the micrometer, which then makes
1068; then, by the following analogy, we fhall obtain the number of times the microfcope magnifies the
diameter of the object; fay, as 8400, the diftance of
the eye from the image of the object, is to 800, the
diftance of the eye from the object; fo is 1068, double
the meafure found on the micrometer, to 5563, or the

Microme-
ter.

number of times the microfcope magnifies the diameter of the object. By working in this manner, the
magnifying power of each lens ufed with the compound
microfcope may be eafily found, though the refult will
be different in different compound microfcopes, varying according to the combination of the lenfes, their
diftance from the object and one another, &c.

Having difcovered the magnifying power of the microfcope, with the different object-lenfes that are ufed
therewith, our next fubject is to find out the real fize
of the objects themfelves, and their different parts:
this is eafily effected, by finding how many revolutions
of the micrometer-fcrew anfwer to a known meafure
on the fectoral fcale or other object placed on the
ftage; from the number thus found, a table fhould be
conftructed, expreffing the value of the different revolutions of the micrometer with that object lens, by
which the primary number was obtained. Similar
tables muft be conftructed for each object lens. By a
fet of tables of this kind, the obferver may readily
find the meafure of any object he is examining; for
he has only to make the needle point traverfe over
this object, and obferve the number of revolutions the
fcrew has made in its paffage, and then look into his
table for the real meafure which correfponds to this
number of revolutions, which is the meafure required.

Mr Coventry of Southwark has favoured us with
the defcription of a micrometer of his own invention;
the fcale of which, for minutenefs, furpaffes every
inftrument of the kind of which we have any knowledge, and of which, indeed, we could fcarcely have
formed a conception, had he not indulged us with feveral of thefe inftruments, graduated as underneath.

The micrometer is compofed of glafs, ivory, filver,
&c. on which are drawn parallel lines from the 10th
to the 10,000th part of an inch. But an inftrument
thus divided, he obferves, is more for curiofity than
ufe; but one of thofe which Mr Coventry has fent us
is divided into fquares, fo fmall that fixteen million of
them are contained on the furface of one fquare inch,
each fquare appearing under the microfcope true and
diftinct; and though fo fmall, it is a fact, that animalcula are found which may be contained in one of
thefe fquares.

The ufe of micrometers, when applied to microfcopes, is to meafure the natural fize of the object,
and how much that object is magnified. To afcertain
the real fize of an object in the fingle microfcope, nothing more is required than to lay it on the micrometer, and adjuft it to the focus of the magnifier, no-
ticing how many divifions of the micrometer it covers.
Suppofe the parallel lines of the micrometer to be the
1000th of an inch, and the object covers two divifions;
its real fize is 500ths of an inch; if five, 200ths, and
fo on.

But to find how much the object is magnified,
is not mathematically determined fo eafy by the fingle
as by the compound microfcope: but the following fimple method (fays Mr Coventry) I have generally adopted, and think it tolerably accurate.
Adjuft a micrometer under the microfcope a, fay the
100th of an inch of divifions, with a fmall object on
it; if fquare, the better: notice how many divifions one
fide of the object covers, fuppofe 10: then cut a piece

4 U 2

of

Microme-
ter.

of white paper something larger than the magnified
appearance of the object: then fix one eye on the object through the microscope, and the other at the same
time on the paper, lowering it down till the object
and the paper appear level and distinct: then cut the
paper till it appear exactly the size of the magnified
object; the paper being then measured, suppose an
inch square: Now, as the object under the magnifier,
which appeared to be one inch square, was in reality
only ten hundredths, or the tenth of an inch, the experiment proves that it is magnified ten times in length,
one hundred times in superfices, and one thousand
times in cube, which is the magnifying power of the
glass; and, in the same manner, a table may be made
of the power of all the other glasses.

In using the compound microscope, the real size of
the object is found by the same method as in the single:
but to demonstrate the magnifying power of each glass
to greater certainty, adopt the following method.—
Lay a two-feet rule on the stage, and a micrometer
level with its surface (an inch suppose, divided into
100 parts); with one eye see how many of those parts
are contained in the field of the microscope, (suppose
50); and with the other, at the same time, look for
the circle of light in the field of the microscope, which
with a little practice will soon appear distinct; mark
how much of the rule is intersected by the circle of
light, which will be half the diameter of the field.
Suppose eight inches; consequently the whole diameter will be sixteen. Now, as the real size of the field,
by the micrometers, appeared to be only 50 hundredths,
or half an inch, and as half an inch is only one 32d
part of 16 inches, it shows the magnifying power of the
glass to be 32 times in length, 1024 superfices, and
32,768 cube (a).

Another way of finding the magnifying power of
compound microscopes, is by using two micrometers
of the same divisions; one adjusted under the magnifier, the other fixed in the body of the microscope in
the focus of the eye-glass. Notice how many divisions of the micrometer in the body are seen in one

division of the micrometer under the magnifier, which
again must be multiplied by the power of the eye-glass.
Example: Ten divisions of the micrometer in the body are contained in one division under the magnifier;
so far the power is increased ten times; now, if the
eye-glass be one inch focus, such glass will of itself magnify about eight times in length, which,
with the ten times magnified before, will be eight
times ten, or 80 times in length, 6403 superfices,
and 512,000 cube.

"If (says Mr Coventry) these micrometers are employed in the solar microscope, they divide the object
into squares on the screen in such a manner as to render it extremely easy to make a drawing of it. And
(says he) I apprehend they may be employed to great
advantage with such a microscope as Mr Adams's Lucernal; because this instrument may be used either by
day or night, or in any place, and gives the actual
magnifying power without calculation."

The only one with which we have been favoured by Mr
Coventry contains six micrometers, two on ivory and
four on glass. One of those on ivory is an inch divided into one hundred parts, every 10th line longer
than the intermediate ones, and every tenth longer still,
for the greater ease in counting the divisions under the
microscope, and is generally used in measuring the
magnifying power of microscopes. The other ivory
one is divided into squares of the 10th and 100th of
an inch, and is commonly employed in measuring
opaque objects.

Those made of glass are for transparent objects,
which, when laid on them, show their natural line.—
That marked on the brass 100, are squares divided to
the 100th of an inch; that marked 1000 are parallel
lines forming nine divisions, each division the 1000th
of an inch; the middle division is again divided into
5, making divisions to the 5000th of an inch. That
marked 10,000 is divided in the same manner, with
the middle division divided into 10, making the
10,000th of an inch. Example:

The glass micrometer without any mark is also divided, the outside lines into 100th, the next into
100th, and the inside lines into the 4000th of an
inch: these are again crossed with an equal number of

lines in the same manner, making squares of the
100th, 1000th, and 4000th of an inch, thus demonstra-
ting each other's size. The middle square of the 100th
of an inch (fee fig. 26.) is divided into sixteen squares;

new

(a) It will be necessary, for great accuracy, as well as for comparative observations, that the two-feet rule
should always be placed at a certain distance from the eye: eight inches would, in general, be a proper distance.

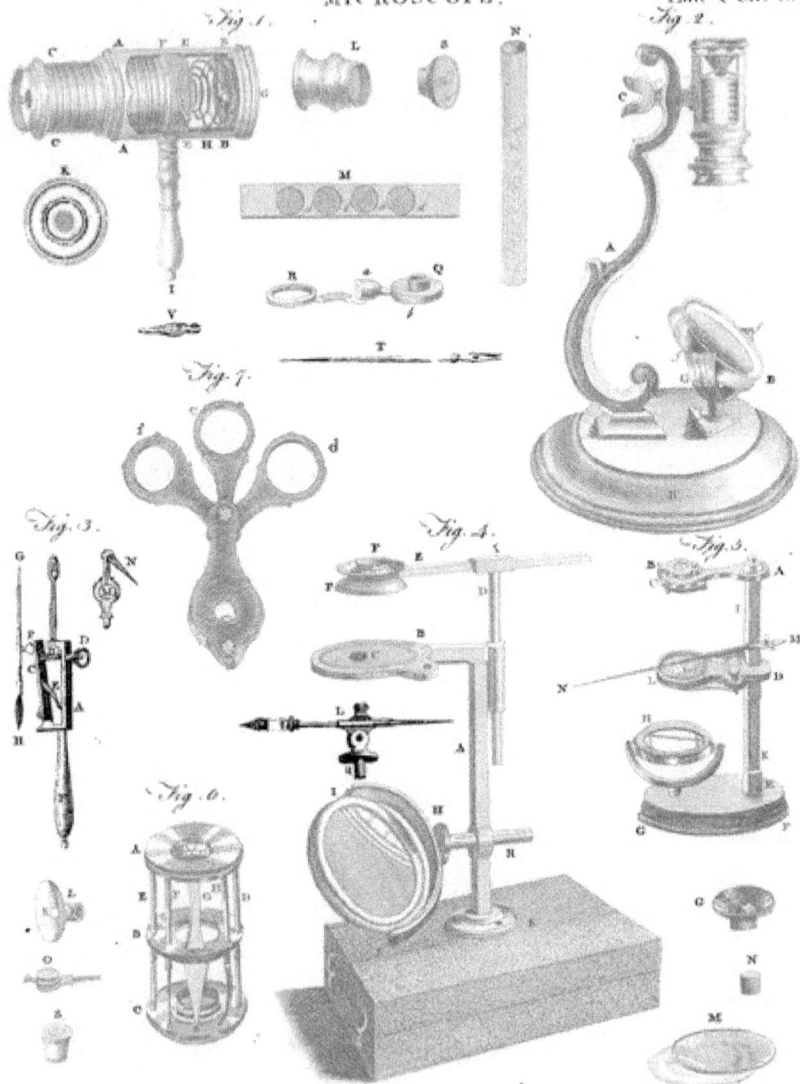

MICROSCOPE.

Plate CCXCVII

Fig. 1.

Fig. 2.

Fig. 7.

Fig. 3.

Fig. 4.

Fig. 5.

Fig. 6.

Microus, now as those squares in the length of an inch, multiplied by 3500, gives one million in an inch surface; by the same rule, one of those squares divided into 16 must be the sixteen-millionth part of an inch surface. See fig. 2, which is a diminished view of the apparent surface exhibited under the magnifier of Wilson's microscope. In viewing the smallest bees, Mr Co. ...

MICROPUS, ...: A genus of the polygamia necessaria order, belonging to the syngenesia class of plants; and in the natural method ranking under the 49th order, Compositæ. The receptacle is paleaceous; there is no pappus; the calyx is calculated; ...

MICROSCOPE, an optical instrument, consisting of lenses, or mirrors, by means of which small objects appear larger than they do to the naked eye. Single microscopes consist of a single lens or mirror; or if more lenses or mirrors be made use of, they only serve to throw light upon the object, but do not contribute to enlarge the image of it. Double or compound microscopes are those in which the image of an object is composed by means of more lenses or mirrors than one.

For the principles on which the construction of microscopes depends, see Optics. In the present article, it is intended to describe the finished instrument, with all its varied apparatus, according to the latest improvements; and to illustrate by proper details its uses and importance.

I. Of Single Microscopes.

The famous microscopes made use of by Mr Leewenhoeck, were all, as Mr Baker assures us, of the single kind, and the construction of them was the most simple possible; each consisting only of a single lens set between two plates of silver, perforated with a small hole, with a moveable pin before it to place the object on and adjust it to the eye of the beholder. He informs us also, that lenses only, and not globules, were used in every one of these microscopes.

1. The single microscope now most generally known and used is that called Wilson's Pocket Microscope. The body is made of brass, ivory, or silver, and is represented by AA, BB. CC is a long fine-threaded male screw that turns into the body of the microscope; D a convex glass at the end of the screw. Two concave round pieces of thin brass, with holes of different diameters in the middle of them, are placed to cover the abovementioned glass, and thereby diminish the aperture when the greatest magnifiers are employed. EE, three thin plates of brass within the body of the microscope; one of which is bent semicircularly in the middle, so as to form an arched cavity for the reception of a tube of glass, the use of the other two being to receive and hold the tube between them. F, a piece of wood or ivory, arched in the manner of the semicircular plate, and cemented to it. G, the other end of the body of the microscope, where a hollow female screw is adapted to receive the different magnifiers. H, is a spiral spring of steel, between the end G and the plates of brass, intended to keep the screw CC. I, is a small turned handle, for the better holding of the instrument, to screw on or off at pleasure.

To this microscope belong six or seven magnifying glasses: six of them are set to silver, brass, or ivory, as in the figure K; and marked 1, 2, 3, 4, 5, 6, the lowest numbers being the greatest magnifiers. L, is the seventh magnifier, set in the manner of a little barrel, to be held in the hand for the viewing of any larger object. M, is a flat slip of ivory, called a slider, with four round holes through it, wherein to place objects between two pieces of glass or Muscovy talc, as they appear at d d d d. Six such sliders, and one of brass, are usually sold with this microscope, some with objects placed in them, and others empty for viewing any thing that may offer: but whoever pleases to make a collection, may have as many as he desires. The brass slider is to confine any small object, that it may be viewed without crushing or destroying it. N, is a tube of glass contrived to confine living objects, such as frogs, fishes, &c. in order to discover the circulation of the blood. All these are contained in a little neat box of fish-skin or mahogany, very convenient for carrying in the pocket.

When an object is to be viewed, thrust the ivory slider, in which the said object is placed, between the two last brass plates EE; observing always to put that side of the slider where the brass rings are farthest from the eye. Then screw on the magnifying glass you intend to use, at the end of the instrument G; and looking through it against the light, turn the long screw CC, till your object be brought to suit your eye; which will be known by its appearing perfectly distinct and clear. It is most proper to look at it first through a magnifier that can show the whole at once, and afterwards to inspect the several parts more particularly with one of the greatest magnifiers; for thus you will gain a true idea of the whole, and of all its parts. And though the greatest magnifiers can show but a minute portion of any object at once, such as the claw of a flea, the horn of a louse, or the like; yet by gently moving the slider which contains the object, the eye will gradually examine it all over.

As objects must be brought very near the glasses when the greatest magnifiers are made use of, be careful not to scratch them by rubbing the slider against them as you move it in or out. A few turns of the screw CC will easily prevent this mischief, by giving them room enough. You may change the objects in your sliders for any others you think proper, by taking out the brass rings with the point of a penknife; the talcs will then fall out, if you but turn the sliders and ...

Microscope, and after putting what you please between them, by replacing the brass rings you will fasten them as they were before. It is proper to have some sliders furnished with talc, but without any object between them, to be always in readiness for the examination of fluids, salts, sands, powders, the farina of flowers, or any other casual objects of such sort as need only be applied to the outside of the talc.

The circulation of the blood may be easiest seen in the tails or fins of fishes, in the fine membrane between a frog's toes, or back of all in the tail of a water-newt. If your object be a small fish, place it within the tube N, and spread its tail or fin along the side thereof: if a frog, choose such an one as can but just be got into your tube; and, with a pen, or small stick, expand the transparent membrane between the toes of the frog's hind foot as much as you can. When your object is so adjusted that no part of it can intercept the light from the place you intend to view, unscrew the long screw CC, and thrust your tube into the arched cavity, quite through the body of the microscope; then screw it to the true focal distance, and you will see the blood passing along its vessels with a rapid motion, and in a most surprising manner.

The third or fourth magnifiers may be used for frogs or fishes; but for the tails of water-newts, the fifth or sixth will do; because the globules of their blood are twice as large as those of frogs or fish. The first or second magnifier cannot well be employed for this purpose; because the thickness of the tube in which the object lies, will scarce admit its being brought so near as the focal distance of the magnifier.

An apparatus for the purpose of viewing opaque objects generally accompanies this microscope; and which consists of the following parts. A brass arm QR, which is screwed at Q, upon the body of the microscope at G. Into the round hole R, any of the magnifiers suitable to the object to be viewed are to be screwed; and under it, in the same ring, the concave polished silver speculum S. Through a small aperture in the body of the microscope under the brass plates EE, is to slide the long wire with the forceps T: This wire is pointed at one of its ends; and so, that either the points or forceps may be used for the objects as may be necessary. It is easy to conceive, therefore, that the arm at R, which turns by a twofold joint at a and b, may be brought with its magnifier over the object, the light reflected upon it by the application of the speculum, and the true focus obtained by turning of the male screw CC as before directed.—As objects are sometimes not well fixed for view, either by the forceps or point, the small piece shown at N is added, and in such cases answers better: it screws over the point of T; it contains a small round piece of ivory, blackened on one side, and left white upon the other as a contrast to coloured objects, and by a small piece of watch-spring fastens down the objects upon the ivory.

2. *Single Microscope by reflection.* In fig. 2. A is a scroll of brass fixed upright upon a round wooden base B, or mahogany drawer or case, so as to stand perfectly firm and steady. C is a brass screw, that passes through a hole in the upper limb of the scroll into the side of the microscope D, and screws it fast to the said scroll. E is a concave speculum set in a box of brass, which hangs in the arch G by two small screws ff, that screw into the opposite sides thereof. At the bottom of this arch is a pin of the same metal, exactly fitted to a hole b in the wooden pedestal, made for the reception of the pin. As the arch turns on this pin, and the speculum turns on the end of the arch, it may, by this twofold motion, be easily adjusted in such a manner as to reflect the light of the sun, of the sky, or of a candle, directly upwards through the microscope that is fixed perpendicularly over it; and by so doing may be made to answer many purposes of the large double reflecting microscope. The body of the microscope may also be fixed horizontally, and objects viewed in that position by any light you choose; which is an advantage the common double reflecting microscope has not. It may also be rendered further useful by means of a slip of glass; one end of which being thrust through between the plates where the sliders go, and the other extending to some distance, such objects may be placed thereon as cannot be applied in the sliders; and then, having a limb of brass that may fasten to the body of the microscope, and extend over the projecting glass a hollow ring wherein to screw the magnifiers, all sorts of subjects may be examined with great convenience, if a hole be made in the pedestal, to place the speculum exactly underneath, and thereby throw up the rays of light. The pocket microscope, thus mounted, says Mr Baker, " is as easy and pleasant in its use; as fit for the most curious examination of the animalcules and salts in fluids, of the farina in vegetables, and of the circulation in small animals; in short, is as likely to make considerable discoveries in objects that have some degree of transparency, as any microscope I have ever seen or heard of."

The brass scroll A is now generally made to unscrew into three parts, and pack with the microscope and apparatus into the drawer of a mahogany pocket-case, upon the lid of which the scroll is made to fix when in use.

The opaque apparatus also, as above described, is applicable this way by reflection. It only consists in turning the arm R (fig. 1.), with the magnifier over the concave speculum below (fig. 2.), or to receive the light as reflected obliquely from it; the silver speculum screwed into R will then reflect the light, which it receives from the glass speculum, strongly upon the object that is applied upon the wire T underneath.

This microscope, however, is not upon the most convenient construction, in comparison with others now made: it has been esteemed for many years past from its popular name, and recommendation by its makers. Its portability is certainly a great advantage in its favour; but in most respects it is superseded by the microscopes hereafter described.

3. *Microscope for Opaque Objects, called the Single Opaque Microscope.* This microscope remedies the inconvenience of having the dark side of an object next the eye, which formerly was an unsurmountable objection to the making observations on opaque objects with any considerable degree of exactness or satisfaction: for, in all other contrivances commonly known, the nearness of the instrument to the object (when glasses that magnify much are used) unavoidably overshadows it so much, that its appearance is rendered obscure and indistinct. And, notwithstanding ways have

Fig. 5.

Microscope have been tried to point light upon an object, from the sun or a candle, by a convex glass placed on the side thereof, the rays from either can be thrown upon it in such an acute angle only, that they serve to give a confused glare, but are insufficient to afford a clear and perfect view of the object. But this microscope, by means of a concave speculum of silver highly polished, in whose centre a magnifying lens is placed, such a strong and direct light is reflected upon the object, that it may be examined with all imaginable ease and pleasure. The several parts of this instrument, made either of brass or silver, are as follow.

Through the first side A, passes a fine screw B, the other end of which is fastened to the moveable side C. D is a nut applied to this screw, by the turning of which the two sides A and C are gradually brought together. E is a spring of steel that separates the two sides when the nut is unscrewed. F is a piece of brass, turning round in a socket, whence proceeds a small spring tube moving upon a rivet; through which tube there runs a steel wire, one end whereof terminates in a sharp point G, and the other with a pair of plyers H fastened to it. The point and plyers are to thrust into, or take up and hold, any insect or object; and either of them may be turned upwards, as best suits the purpose. I is a ring of brass, with a female screw within it, mounted on an upright piece of the same metal; which turns round on a rivet, that it may be set at a due distance when the least magnifiers are employed. This ring receives the screws of all the magnifiers. K is a concave speculum of silver, polished as bright as possible; in the centre of which is placed a double convex lens, with a proper aperture to look through it. On the back of this speculum a male screw L is made to fit the brass ring I, to screw into it at pleasure. There are four of these concave speculæ of different depths, adapted to four glasses of different magnifying powers, to be used as the objects to be examined may require. The greatest magnifiers have the least apertures. M, is a round object-plate, one side of which is white and the other black: The intention of this is to render objects the more visible, by placing them, if black, on the white side, or, if white, on the black side. A steel spring N turns down on each side to make any object fall; and issuing from the object-plate is a hollow pipe to screw it on the needle's point G, O, is a small box of brass, with a glass on each side, contrived to confine any living object, in order to examine it; this also has a pipe to screw upon the end of the needle G. P, is a turned handle of wood, to screw into the instrument when it is made use of. Q, a pair of brass pliers to take up any object, or manage it with conveniency. R, is a soft hair-brush for cleaning the glasses, &c. S, is a small ivory box for talc, to be placed, when wanted, in the small brass-box O.

When you would view any object with this microscope, screw the speculum, with the magnifier you think proper to use, into the brass ring I. Place your object, either on the needle G in the pliers H, on the object-plate M, or in the hollow brass box O, as may be most convenient; then holding up your instrument by the handle P, look against the light through the magnifying lens; and by means of the nut D, together with the motion of the needle, by managing its lower

end, the object may be turned about, raised, or the Microscope pressed, brought nearer the glass, or removed further from it, till you find the true focal distance, and the light be seen strongly reflected from the speculum upon the object, by which means it will be shown in a manner surprisingly distinct and clear; and for this purpose the light of the sky or of a candle will answer very well. Transparent objects may also be viewed by this microscope; only observing, that when such come under examination, it will not always be proper to throw on them the light reflected from the speculum; for the light transmitted through them, meeting the reflected light, may together produce too great a glare. A little practice, however, will show how to regulate both lights in a proper manner.

4. Ellis's single and aquatic Microscope. Fig. 4. represents a very convenient and useful microscope, contrived by Mr John Ellis, author of An Essay upon Corallines, &c. To practical botanists, observers of animalcula, &c. it possesses many advantages above those just described. It is portable, simple in its construction, expeditious, and commodious in use. K, represents the box containing the whole apparatus: it is generally made of fish-skin; and on the top there is a female screw; and on the top there is a female screw, for receiving the screw that is at the bottom of the pillar A; this is a pillar of brass, and is screwed on the top of the box. D, is a brass pin which fits into the pillar; on the top of this pin is a hollow socket to receive the arm which carries the magnifiers; the pin is to be moved up and down, in order to adjust the lenses to their focal or proper distance from the object. [N. B. In the representations of this microscope, the pin D is delineated as passing through a socket at one side of the pillar A; whereas it is usual at present to make it pass down a hole bored through the middle of the pillar.] E, the bar which carries the magnifying lens; it fits into the socket X, which is at the top of the pin or pillar D. This arm may be moved backwards and forwards in the socket X, and sideways by the pin D; so that the magnifier, which is screwed into the ring at the end E of this bar, may be easily made to traverse over any part of the object that lies on the stage or plate B. FF is a polished silver speculum, with a magnifying lens placed at the centre thereof, which is perforated for this purpose. The silver speculum screws into the arm E, as at F. G, another speculum, with its lens, which is of a different magnifying power from the former. H, the semicircle which supports the mirror I; the pin R, affixed to the semicircle H, passes through the hole which is towards the bottom of the pillar A. B, the stage, or the plate, on which the objects are to be placed; it fits into the small dove-tailed arm which is at the upper end of the pillar DA. C, a plane glass, with a small piece of black silk stuck on it; this glass is to lay in a groove made in the stage B. M, a hollow glass to be laid occasionally on the stage instead of the plane glass C. L, a pair of nippers. These are fixed to the stage by the pin at bottom; the steel wire of these nippers slides backwards and forwards in the socket, and this socket is moveable upwards and downwards by means of the joint, so that the position of the object may be varied at pleasure. The object may be fixed in the nippers, stuck on the point, or affixed, by a little gum-water, &c. to the

may

Microscope ivory cylinder N, which occasionally screws to the point of the nippers.

To use this microscope: Take all the parts of the apparatus out of the box; then begin by screwing the pillar A to the cover thereof; pass the pin R of the semicircle which carries the mirror thro' the hole that is near the bottom of the pillar A; push the finger into the dove-tail at B; slide the pin into the pillar (see the N. B. above); then pass the bar E through the socket which is at the top of the pin D, and screw one of the magnifying lenses into the ring at F. The microscope is now ready for use; and though the enumeration of the articles may lead the reader to imagine the instrument to be of a complex nature, we can safely affirm that he will find it otherwise. The instrument has this peculiar advantage, that it is difficult to put any of the pieces in a place which is appropriated to another. Let the object be now placed either on the stage or in the nippers L, and in such manner that it may be as nearly as possible over the centre of the stage; bring the speculum F over the part you mean to observe; then throw as much light on the speculum as you can, by means of the mirror I, and the double motion of which it is capable; the light received on the speculum is reflected by it on the object. The distance of the lens F from the object is regulated by moving the pin D up and down, until a distinct view of it is obtained. The best rule is, to place the lens beyond its focal distance from the object, and then gradually to slide it down till the object appears sharp and well defined. The adjustment of the lenses to their focus, and the distribution of the light on the object, are what require the most attention: on the first the distinctness of the vision depends; the pleasure arising from a clear view of the parts under observation is due to the modification of the light. No precise rule can be given for obtaining accurately these points; it is from practice alone that ready habits of obtaining these necessary properties can be acquired, and with the assistance of this no difficulty will be found.

5. A very simple and convenient microscope for botanical and other purposes, though inferior in many respects to that of Mr Ellis, was contrived by the late ingenious Mr Benjamin Martin, and is represented at fig. 5. where A B represents a small arm supporting two or more magnifiers, one fixed to the upper part as at B, the other to the lower part of the arm at C; these may be used separately or combined together. The arm A B is supported by the square pillar IK, the lower end of which fits into the socket E of the foot F G; the stage DL is made to slide up and down the square pillar; H, a concave mirror for reflecting light on the object.—To use this microscope, place the object on the stage, reflect the light on it from the concave mirror, and regulate it to the focus, by moving the stage nearer to or farther from the lens at B. The ivory sliders pass through the stage; other objects may be fixed in the nippers MN, and then brought under the eye-glasses; or they may be laid on one of the glasses which fit the stage. The apparatus to this instrument consists of three ivory sliders, a pair of nippers; a pair of forceps; a flat glass and a concave ditto, both fitted to the stage.

The two last microscopes are frequently fitted up

N° 115.

with a toothed rack and pinion, for the more ready Microscope adjustment of the glasses to their proper focus.

6. *Withering's portable Botanic Microscope.* Fig. 6. represents a small botanical microscope contrived by Dr Withering, and described by him in his *Botanical Arrangements*. It consists of three brass plates, A, B, C, which are parallel to each other; the wires D and E are rivetted into the upper and lower plates, which are by this means united to each other; the middle plate or stage is moveable on the aforesaid wires by two little sockets which are fixed to it. The two upper plates each contain a magnifying lens, but of different powers; one of these confines and keeps in their places the fine point F, the forceps G, and the small knife H.—To use this instrument, unscrew the upper lens, and take out the point, the knife, and the forceps; then screw the lens on again, place the object on the stage, and then move it up or down till you have gained a distinct view of the object, as one lens is made of a shorter focus than the other; and spare lenses of a still deeper focus may be had if required. This little microscope is the most portable of any. Its principal merit is its simplicity.

7. *Botanical Lenses or Magnifiers.* The haste with which botanists, &c. have frequently occasion to view objects, renders an extempore pocket-glass indispensably necessary. The most convenient of any yet constructed, appears to be that contrived, in regard to the form of the mounting, by the late Mr Benjamin Martin; and is what he called a *Hand Magniscope*, because it is well adapted for viewing all the larger sort of small objects universally; and by only three lenses it has seven different magnifying powers.

Fig. 7. represents the case with the three frames and lenses, which are usually of 2, 1¼, and 1 inches focus: they all turn over each other, and shut into the case, and are turned out at pleasure.

The three lenses singly, afford three magnifying powers; and by combining two and two, we make three more: for d with e makes one, d with f another, and e with f a third; which, with the three singly, make six; and lastly, all three combined together make seven; so that upon the whole, there are seven powers of magnifying with these glasses only.

When the three lenses are combined, it is better to turn them in, and look through them by the small apertures in the sides of the case. The eye in this case is excluded from extra light; the aberration of the superfluous rays through the glasses is cut off; and the eye coincides more exactly with the common axis of the lenses.

A very useful and easy kind of microscope (described by Joblot, and which has been long in use), adapted chiefly for viewing, and combining at the same time, any living insects, small animals, &c. is shown at fig. 8. where A represents a glass tube, about 1½ inches diameter, and 2 inches high. B, a case of brass or wood, containing a sliding tube, with two or three magnifying glasses that may be used either separately or combined. In the inside, at the bottom, is a piece of ivory, black and white on opposite sides, that is occasionally removed, and admits a point to be screwed into the centre. The cap unscrews at D, to admit the placing of the objects the proper distance of the glasses from

Plate CCXCVIII

Fig. 8.

Fig. 9.

Fig. 10.

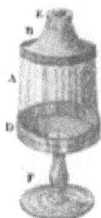

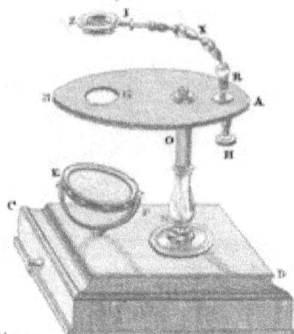

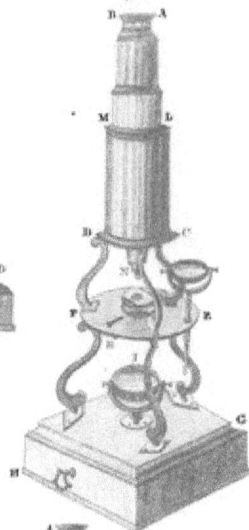

Fig. 11.

Fig. 12.

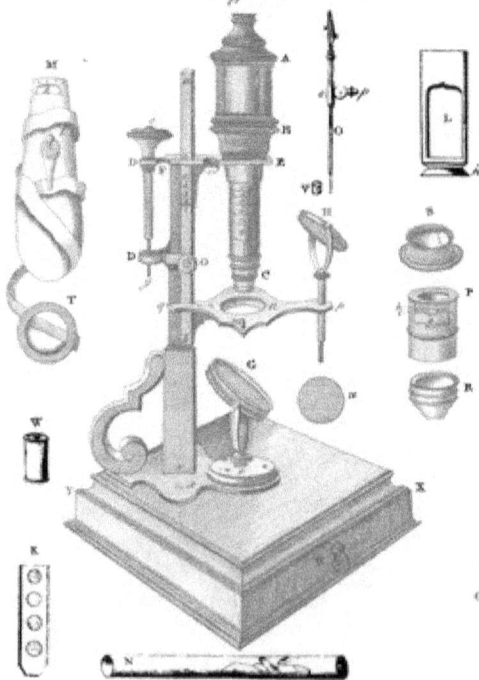

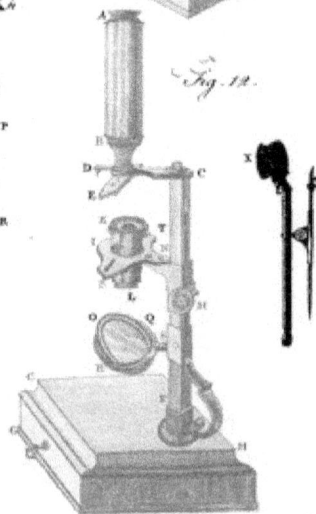

Microscope from the object is regulated by pulling up or down the brass tube E at top containing the eye-glasses.

This microscope is particularly useful for exhibiting the well known curious *animalcula imperatilia*, vulgarly called the *diamond beetle*, to the greatest advantage; for which, as well as for other objects, a glass bottom, and a polished reflector at the top, are often applied, to condense the light upon the object. In this case, the stand and brass-bottom F, as shown in the figure, are taken away by unscrewing.

9. *Mr Lyonet's Single Anatomical Dissecting Microscope.* Fig. 9. represents a curious and extremely useful microscope, invented by that gentleman for the purpose of minute dissections, and microscopic preparations. This instrument must be truly useful to amateurs of the minutiæ of insects, &c. being the best adapted of any for the purposes of dissection. With this instrument Mr Lyonet made his very curious microscopical dissection of the *chenille du saule*, as related in his *Traité Anatomique de la chenille qui ronge le bois de saule*, 4to.

AB is the anatomical table, which is supported by a pillar NO; this is screwed on the foot CD. The table AB is prevented from turning round by means of two steady pins. In this table or board there is a hole G, which is exactly over the centre of the mirror EF, that is to reflect the light on the object; the hole G is designed to receive a flat or concave glass, on which the objects for examination are to be placed.

RXZ is an arm formed of several balls and sockets, by which means it may be moved in every possible direction; it is fixed to the board by means of the screw H. The last arm IZ has a female screw, into which a magnifier may be screwed as at Z. By means of the screw H, a small motion may be occasionally given to the arm IZ, for adjusting the lens with accuracy to its focal distance from the object.

Another chain of balls is sometimes used, carrying a lens to throw light upon the object; the mirror is likewise so mounted, as to be taken from its place at K, and fitted on a clamp, by which it may be fixed to any part of the table AB.

To *use the Dissecting Tables*—Let the operator sit with his left side near a light window; the instrument being placed on a firm table, the side DH towards the stomach, the observations should be made with the left eye. In dissecting, the two elbows are to be supported by the table on which the instrument rests, the hands resting against the board AB; and in order to give it greater stability (as a small shake, though imperceptible to the naked eye, is very visible in the microscope), the dissecting instruments are to be held one in each hand, between the thumb and two fore-fingers.

II. *Of Double Microscopes, commonly called Compound Microscopes.*

Double microscopes are so called, from being a combination of two or more lenses.

The particular and chief advantages which the compound microscopes have over the single, are, that the objects are represented under a larger field of view, and with a greater amplification of reflected light.

1. *Culpeper's Microscope.* The compound microscope, originally contrived by Mr Culpeper, is represented at

fig. 10. It consists of a large external brass body A, B, C, D, supported upon three scrolls, which are fixed to the stage EF; the stage is supported by three larger scrolls, that are screwed to the mahogany pedestal GH. There is a drawer in the pedestal, which holds the apparatus. The concave mirror I is fitted to a socket in the centre of the pedestal. The lower part LMCD of the body forms an exterior tube, into which the upper part of the body ABLM slides, and may be moved up or down, so as to bring the magnifiers, which are screwed on at N, nearer to or further from the object.

To *use* this microscope: Screw one of the buttons, which contains a magnifying lens, to the end N of the body; place the slider, with the objects, between the plates of the slider-holder. Then, to attain distinct vision, and a pleasing view of the object, adjust the body to the focus of the lens you are using, by moving the upper part gently up and down, and regulate the light by the concave mirror.

For opaque objects, two additional pieces must be used. The first is a cylindrical tube of brass (represented at L, fig. 11.), which fits on the cylindrical part at N of the body. The second piece is the concave speculum, lum L; this is to be screwed to the lower end of the aforesaid tube : the upper edge of this tube should be made to coincide with the line which has the same number affixed to it as to the magnifier you are using; or, if you are making use of the magnifier marked 5, slide the tube to the circular line on the tube N that is marked also with N 5. The tube-holder should be removed when you are going to view opaque objects, and a plane glass should be placed on the stage in its stead to receive the object ; or it may be placed in the nippers, the pin of which fits into the hole in the stage.

The apparatus belonging to this microscope consists of the following particulars; viz. Five magnifiers, each fitted in a brass button; one of these is seen at N, fig. 10. Six ivory sliders, five of them with objects. A brass tube, to hold the concave speculum. The concave speculum in a brass box. A fish pan. A set of glass tubes. A flat glass fitted to the stage. A concave glass fitted to the stage. A pair of forceps. A steel wire, with a pair of nippers at one end and a point at the other. A small ivory cylinder, to fit on the pointed end of the aforesaid nippers. A convex lens, moveable in a brass semicircle ; this is affixed to a long brass pin, which fits into a hole on the stage.

The construction of the foregoing microscope is very simple, and it is easy in use; but the advantages of the stage and mirror are too much confined for an extensive application and management of all kinds of objects. Its greatest recommendation is its cheapness ; and to those who are desirous of having a compound microscope at a low price, it may be acceptable.

2. *Cuff's Microscope.* The improved microscope next in order is that of Mr Cuff. Besides remedying the disadvantages above mentioned, it contains the addition of an adjusting screw, which is a considerable improvement, and highly necessary to the examination of objects under the best defined appearance from the glasses. It is represented at fig. 11. with the apparatus that usually accompanies it. A, B, C, shows the body of this microscope; which contains an eye-glass at A, a broad lens at B, and

4 X

Microscope a magnifier which is screwed on at C. The body is supported by the arm D E, from which it may be removed at pleasure. The arm D E is fixed on the sliding-bar F, and may be raised or depressed to any height within its limits. The main pillar *a b* is fixed in the box *b e*; and by means of the brass foot *d* is screwed to the mahogany pedestal X Y, in which is a drawer containing all the apparatus. O, is a milled-headed screw, to tighten the bar F when the adjusting screw *c g* is used. *p p* Is the stage, or plate, which carries the objects; it has a hole at the centre *e*. G a concave mirror, that may be turned in any direction, to reflect the light of a candle, or the sky, upon the object.

To use this microscope: Screw the magnifier you intend to use to the end C of the body, place the slider-holder P in the hole *n*, and the slider with the object between the plates of the slider-holder; set the upper edge of the bar D E to coincide with the divisions which correspond to the magnifier you have in use, and pinch it by the milled nut; now reflect a proper quantity of light upon the object, by means of the concave mirror G, and regulate the body exactly to the eye and the focus of the glasses by the adjusting screw *c g*.

To view opaque objects, take away the slider-holder P, and place the object on a flat glass under the centre of the body, or on one end of the jointed nippers *o p*. Then screw the silver concave speculum *b* to the end of the cylinder L, and slide this cylinder on the lower part of the body, so that the upper edge thereof may coincide with the line which has the same mark with the magnifier that is then used; reflect the light from the concave mirror G to the silver speculum, from which it will again be reflected on the object. The glasses are to be adjusted to their focal distance as before directed.

The apparatus consists of a convex lens H, to collect the rays of light from the sun or a candle, and condense them on the object. L a cylindrical tube, open at each side, with a concave speculum screwed to the lower end *b*. P the slider-holder: this consists of a cylindrical tube, in which an inner tube is forced upwards by a spiral spring; it is used to receive an ivory slider K, which is to be slid between the plates *b* and *i*. The cylinder P fits the hole *n* in the stage; and the hollow part at *b* is designed to receive a glass tube. R is a brass cone, to be put under the bottom of the cylinder P, to intercept occasionally some of the rays of light. S a box containing a concave and a flat glass, between which a small living insect may be confined: it is to be placed over the hole *n*. T a flat glass, to lay any occasional object upon; there is also a concave one for fluids. O is a long steel wire, with a small pair of pliers at one end, and a point at the other, designed to stick or hold objects; it slips backwards and forwards in the short tube *e*; the pin *p* fits into the hole of the stage. W a little round ivory box, to hold a supply of tale and rings for the sliders. V a small ivory cylinder, that fits on the pointed end of the steel wire: it is designed for opaque objects. Light-coloured ones are to be stuck upon the dark side, and vice versa. M a fish-pan, whereon to fasten a small fish, to view the circulation of the blood: the tail is to be spread across the oblong hole

at the small end, and tied fast, by means of a ribband fixed thereto; the knob *i* is to be shoved through the slit made in the stage, that the tail may be brought under the magnifier.

5. This microscope has received several material improvements from Sir Martin, Mr Adams, &c. By an alteration, or rather an enlargement, of the body of the tube which contains the eye glasses, and also of the eye-glasses themselves, the field of view is made much larger, the mirror below for reflecting light is made to move upon the same bar with the stage by which means the distance of it from the stage may be very easily and sensibly varied. A condensing glass is applied under the stage in the slider-holder, in order to modify and increase the light that is reflected by the mirror below from the light of a candle or lamp. It is furnished also with two mirrors in one frame, one concave and the other plane, of glass silvered; and by simply unscrewing the body, the instrument, when desired, may be converted into a single microscope. Fig. 2. is a representation of the instrument thus improved; and the following is the description of it, as given by Mr Adams in his Essays.

A B represents the body of the microscope, containing a double eye-glass and a body-glass: it is here shown as screwed to the arm C D, from whence it may be occasionally removed, either for the convenience of packing, or when the instrument is to be used as a single microscope.

The eye glasses and the body glasses are contained in a tube which fits into the exterior tube A B; by pulling out a little this tube when the microscope is in use, the magnifying power of each lens is increased.

The body A B of the microscope is supported by the arm C D; this arm is fixed to the main pillar C F, which is screwed firmly to the mahogany pedestal G H; there is a drawer to this pedestal, which holds the apparatus.

N I S, The plate or stage which carries the slider-holder K L: this stage is moved up or down the pillar C F, by turning the milled nut M; this nut is fixed to a pinion, that works in a toothed rack cut on one side of the pillar. By means of this pinion, the stage may be gradually raised or depressed, and the object adjusted to the focus of the different lenses.

K L is a slider-holder, which fits into a hole that is in the middle of the stage N I S; it is used to confine and guide either the motion of the sliders which contain the objects, or the glass tubes that are designed to confine small fishes for viewing the circulation of the blood. The sliders are to be pushed between the two upper plates, the tubes through the bent plates.

L is a brass tube, to the upper part of which is fixed the condensing lens before spoken of; it fits into the under part of the slider-holder K L, and may be set at different distances from the object, according to its distance from the mirror or the candle.

O is the frame which holds the two reflecting mirrors, one of which is plane, the other concave. These mirrors may be moved in various directions, in order to reflect the light properly, by means of the pivots on which they move, in the semicircle Q S R, and the motion of the semicircle itself on the pin S: the concave mirror generally answers best in the day-time; the plane mirror combines better with the condensing lens.

7

Microel ore lens, and a lamp or candle. At D there is a socket for receiving the pin of the arm Q, (fig. 51.), in which the concave speculum, for reflecting light on opaque objects, is fixed. At S is a hole and flit for receiving either the support L (fig. 51. Pl. ccxl.) or the Lithopan I; when these are used, the slider-holder must be removed. T, a hole to receive the pin of the convex lens M, fig. 51.

To use this microscope: Take it out of the box. Screw the body into the round end of the upper part of the arm C D. Place the lens slider, which contains the magnifiers, into the dove-tailed flit which is on the under side of the aforesaid arm, as seen at E, and slide it forwards until the magnifier you mean to use is under the centre of the body: opposite to each magnifier in this flit there is a notch, and in the dove-tailed part of the arm C D there is a spring, which falls into the above-mentioned notch, and thus makes each magnifier coincide with the centre of the body. Place the ivory sliders you intend to use between the upper glasses of the slider-holder K L, and then reflect as strong a light as you can on the objects by means of one of the mirrors; after that, adjust the object to the focus of the magnifier and your eye, by turning the milled screw M, the motion of which raises and depresses the stage N I S. The degree of light necessary for each object, and the accuracy required in the adjustment of the lenses to their proper focal distance from the object will be easily attained by a little practice.

When opaque objects are to be examined, remove the slider-holder, and place the object on a flat glass, or fix it to the support L, the pin of these fit into the hole on the stage; screw the concave speculum R into the arm Q, fig. 51., and then pass the pin of this arm through the socket D, fig. 51. the light is now to be reflected from the concave mirror to the silver speculum, and from thence down on the object. No exact rule can be given for reflecting the light on the object; we must therefore refer the reader to the manner of all opticals, practice. The speculum must be moved lower or higher, to suit the focus of the different magnifiers and the nature of the object.

The foregoing directions apply equally to the using of this instrument as a single microscope; with this difference only, that the body AB is then removed, and the eye is applied to the upper surface of the arm CD, exactly over the magnifiers.

This microscope is sometimes made with the following alterations, which are supposed to make it still more convenient and useful. The arm CD that carries the body and magnifiers is made both to turn on a pin, and to slide backwards and forwards in a socket at C; so that, instead of moving the objects below on the stage, and disturbing them, the magnifiers are more conveniently brought over any part of the objects as desired. The condensing glass is made larger, and slides upon the square bar CF quite distinct from the stage, like the mirrors below; and it is thereby made useful for any other objects that may be applied on glasses fitted to the stage, as well as those put into the slider-holder K. It is thereby not confined to this flace alone, as in the preceding. When the body AB is taken away, the arm CD may be slipt away from its bar, with the magnifiers, and the forceps, wire, and joint, applied to it; and it thereby

by screws the perpole of a small hand flug's or opaque microscope, for any object occasionally applied to this wire. The magnifiers in the slider E are mounted on a wheel cafe, which perhaps prevents its being in the way to touch as the long slider E before described.— This contrivance is represented at N, fig. 51. Microscope

4. *Martin's New Universal Compound Microscope* — This instrument was originally constructed by the late Mr B. Martin, and intended to comprise all the ufes and advantages of the single, compound, opaque, and aquatic microscopes. The following is a description of it as now made, with a few alterations, chiefly suggested (we are told) by Mr Jones of Holborn.

Fig. 53. is a representation of the instrument placed up for use. A, B, C, D, is the body of the microscope: which consists of four parts, viz. AB the eye-piece, or that containing the eye-glasses, and is screwed into C, which is a moveable or sliding tube on the top; this inner tube contains the body-glass screwed into its lower part. D is the exterior tube or cafe, in which the other slides up and down in an easy and steady manner. This motion of the tube C is useful to increase or decrease the magnifying power of the body-glass when thought necessary, as before mentioned. E is a pipe or snout screwed unto the body of the microscope D, and at its lower part, over the several magnifying lenses hereafter described. FGHI is the square stem of the microscope, upon which the stage It moves in an horizontal position, upwards or downward, by means of the fine rackwork of teeth and pinion. KL is a strong folid joint and pillar, by which the position of the instrument is readily altered from a vertical one to an oblique or to a perfectly horizontal one, as may be required: it is thus well adapted to the rate of the observer either fitting or standing; and as it is very often convenient to view objects by direct transmitted light, when the square stem FI is placed in an horizontal position for this purpose, the mirror T is then to be taken off in order to prevent the obstruction of the rays. M is a circular piece of brass, serving as a bale to the pillar. NOP, the tripod or foot by which the whole body of the microscope is steadily supported; it folds up when packed into the cafe. W is a brass frame, that contains the condensing lens, and acts in combination with the large concave and plane mirrors below at T; the reflected rays from which, either of the common light or of that of a candle or lamp, it agreeably modifies, and makes steady in the field of view.

The particulars of the apparatus to this microscope are as follow: Q is a circular brass box, containing fix magnifiers or object lenses, numbered 1, 2, 3, 4, 5, 6; the digits of which appear feverally through a small round hole in the upper plate of it. To the upper side is fixed a small circle of brafs, by which it is connected with, and screwed into, the round end of the arm abcd, which is a long piece of brass, and moves through either by teeth or pinion, or not, as may be defired, in cd; which is a socket on the upper part of the pillar, and admits, with a motion both eafy and steady, the brafs arm. R is a fixed flage, upon which the objects to be viewed are to be placed: it is firmly fastened to the square pillar, which is moved by the rackwork. In the middle is a large circular hole, for receiving Plate CCXXIX.

4 X 2

Fig. 13.

Fig. 14.

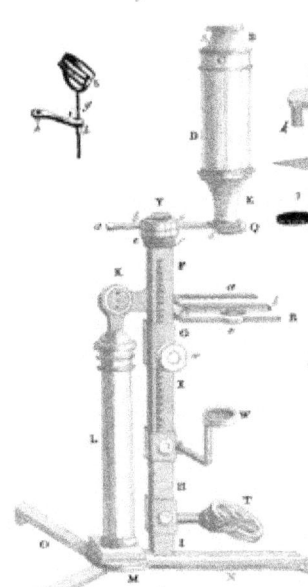

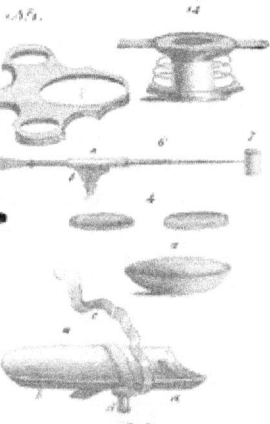

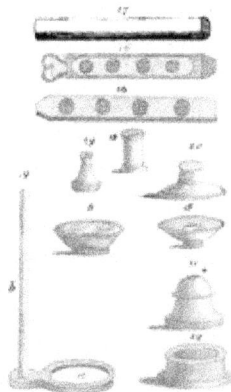

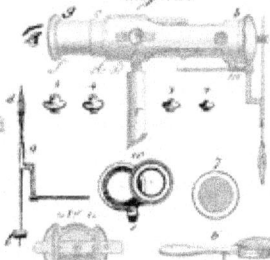

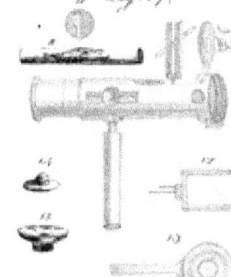

Fig. 16.

Fig. 17.

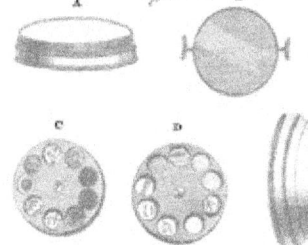

Fig. 20.

Fig. 19.

Fig. 15.

Fig. 18.

Microscope of the objects may be viewed without the least disturbance.

As the brass arm *abcd* may be brought to the height of three or four inches above the stage R; so, by means of the rack-work motion of the stage, a lens of a greater focal distance than the greatest in the wheel Q may be occasionally applied in place of the wheel, and thereby the larger kind of objects viewed; the instrument becoming, in this case, what is called a *megaloscope*.

In viewing moving living objects, or even fixed ones, when nice motions are requisite, a rack-work and pinion is often applied to the arm *abcd*: the arm is cut out with teeth; and the pinion, as shown at Y, is applied to work it. This acts but in one direction; and, in order to produce an equally necessary motion perpendicular to this, rack-work and pinion is applied tangent-wise to the stage, which is thus jointed.

What has been related above respecting the construction of those denominated *pocket microscopes*, in contradistinction to those which are portable, their dimensions, however, have been considerably reduced by opticians, in order to render them fit for the pocket; and as they are for the most part constructed on nearly the same principles as those which have been already described, what has been said will sufficiently instruct our readers in using any pocket microscope whatever. Only it may be observed, that in those reduced instruments, both the field of view and the magnifying power are proportionably diminished.

We shall conclude the account of this sort of microscope with descriptions of a very portable pocket apparatus of microscopic instruments, and of a new microscopic pocket-telescope, both invented by the late Mr B. Martin, and since made by most instrument-makers in London.

The former is represented at fig. 15. It consists of two parts, viz. the body *a b*, and the pedestal *cd*, which is joined by a screw at the part between *b* and *i*. It consists of three cylindric tubes, viz. (1.) the exterior tube, or case *ab*; (2.) a middle tube *cb*; and (3.) the interior tube *fg*.—The middle tube *cd* is the adjuster; and is connected with the outer tube by the rack-work of teeth and pinion, as shown at *e*; by which means it is moved up and down at pleasure through the smallest space; and carries with it the internal tube *fg*. The interior tube *fg* receives on its lower part at *f* the several capsules or boxes 2, 3, 4, 5, (fig. 16.) which contain the object lenses or magnifiers.

The method of using this compound microscope in the perpendicular position, is as follows. The stage n° 1. is put within the exterior tube at *b*. Under the springs are applied the four ivory sliders, which contain a variety of transparent objects; then move the interior tube *fg* up and down with the hand, till you discern the object in the slider, and there let it rest. After this, turn the pinion at *e* very tenderly one way or the other, till you obtain a perfect view of the transparent objects properly illuminated, from a mirror contained in the pedestal or stand *cd*, suspended upon, and moveable about, the points of two screws *(ll)*. N° 6. (fig. 16.) represents a moveable stage, which is placed in the spring socket m. It contains a concave glass, for the reception of animalcules in fluids; and has the advantage of bringing any part into view by moving the handle at n. If living and moving objects are required to be shown, they must be confined in the concave, by putting a glass cover, n° 7. upon the stage; and then a small spider, a louse, flea, bug, &c. may be seen, and the motion or circulation of the blood, &c. observed with surprising distinctness.

To view the circulation of the blood in the most eminent degree, it must be done by placing small frogs, tadpoles, water-newts, fishes, &c. in a tube as represented n° 8. (fig. 17.); which tube is placed in the holes *o* in the opposite sides of the case *ab*, fig. 15. in the lower part.—N° 9. (fig. 16.) is a pair of pincers or pliers *d*, for holding any object; the other end of the steel wire is pointed to receive a piece of ivory *b*, with one end black, and the other white, on which you stick objects of different hues; this also, when used, is placed in the spring socket m.

To use this instrument as a compound opaque, you screw off the body part *ab*, and screw to it the handle *r* (fig. 16.); by this means you may hold the microscope in a horizontal position, as shown in the figure. The silver dish or speculum (which is contained in the bottom or base *b*, fig. 15.), is then screwed on at *b*. N° 9. is placed in the spring-socket *m*, and adjusted backward and forward in *m*, till the reflected light from the speculum falls in a proper manner on the opaque object. Either of the 4 magnifiers, 2, 3, 4, 5, may be used, and brought to a proper focus, as before described, by the teeth and pinion *e* (fig. 15.) If you take off the opaque apparatus, and apply the stage n° 1. (fig. 16.) with an ivory slider, and at the end *b* screw in either of the two lenses, n° 10. (which are distinguished by the name of illuminators), the microscope being held up to the light (and properly adjusted), the whole field of view will be strongly illuminated, and present a most pleasing appearance of any transparent object. These two convex lenses are of different focuses, and are to be used singly or together; n° 10. being the greatest magnifies, will require the object to be strongly illuminated, and of course both the lenses must be used together. By candlelight, this method of viewing transparent objects will prove very entertaining; by screwing the handle *r* into the part *c* of n° 10. it becomes a delightful hand megaloscope for viewing flowers, fossils, shells, &c.; and each lens, as before mentioned, having a different focus, produces two magnifying powers used singly, and when combined a third.

The manner of using this instrument as a single microscope (like Wilson's) is represented in fig. 17. where the bottom or magnifier at each is to be screwed off, and the circular piece n° 11. is screwed in its place. This piece has a spring socket made to receive the slider holder n° 12. N° 13. is a circular piece of brass, with a long shank and spring, and is introduced through the outside tube *ab* at *e*. N° 2, 3, 4, 5, are screwed occasionally in the centre of this piece, and used as single lenses with ivory sliders, &c. N° 14. contains a lens of a great magnifying power, for viewing very minute objects; to render this instrument the most complete single opaque microscope, you have only to screw into n° 13. the silver speculum

n° 15.

Microscope n° 16, which has a small lens set in its centre. The slider-holder n° 12 is taken out of n° 11, and the pincers or nippers *d b*, being detached from the other part of n° 9, are passed through the long spring socket n° 11, and ready to receive any opaque body in the pincers or on the black and white piece of ivory. To the large screw of n° 13. are applied the two lenses n° 10. which make it the completest megalascope that can be desired.

The handle *r* contains the four ivory sliders with objects.

The shagreen case which contains this universal microscope and its apparatus, is six inches long, three inches wide, two inches deep, and weighs together 16 ounces. " Thus (says Mr Martin) so small, so light, so portable, and yet so universally complete, is this pocket microscopic apparatus, that you find nothing material in the large three-pillared microscope, the opaque microscope, Wilson's single microscope, and the aquatic microscope, all together, which you have not in this; beside some very considerable advantages in regard to the field of view, &c. which they have not (A)."

This inventive artist having contrived a construction of the compound microscope so small as to admit of being packed in a common walking cane, thought next of introducing the same instrument into the inside of what he called his *Pocket Three-length drawer Achromatic Telescope.* The same eye-glasses that serve the purpose of a telescope, answer as the compound magnifier, for viewing transparent and opaque objects in a microscope.

Fig. 18, 19, 20. represent the telescope separated by unscrewing it at *u*, in order that the whole of the necessary parts in use may be exhibited. Fig. 19. represents the exterior tube, which is of mahogany, and its rims of brass. It is detached from the rest of the telescope, as not making any part of the microscope. The brass cover *I I*, that shuts up the object-glass of the telescope, is also the box which contains the two-wheel object-frames, and a small plain reflecting mirror.

In fig. 20. A is the cover taken off, by unscrewing the top part: The mirror B is taken out; and also, by unscrewing the bottom part, the two circular wheels, with the objects shown in C and D.

Fig. 18. is a representation of the three internal brass sliding tubes of the telescope, which form the microscopic part. The tubes are to be drawn out as shown in this figure; then, at the lower end of the large tube in the inside, is to be pulled out a short tube *b*, that serves as a kind of stage to hold the wheels with objects, and support the reflecting mirror. This tube is to be partly drawn out, and turned so that the circular hole that is pierced in it may coincide with a similar hole that is cut in the exterior tube.

This tube is represented as drawn out in the figure, and the mirror B placed therein, and the wheel with transparent objects. C (fig. 20.) represents the wheel with transparent objects, and D the wheel with opaque objects. They are both made of ivory; and turn round upon a centre brass pin fit upon the top, which sits upon the edge of the tube; which tube is then to be pushed up into the telescope tube, so that its lower end may rest upon the upper edge of the wheel according to its view at *a* fig. 18.

In viewing the objects, the second brass tube of the telescope must be pushed down, till its milled edge at top falls upon that of the exterior tube; taking care that the circular hole is duly placed to the exterior one. These circular holes are not seen in fig. 18. being supposed in the opposite side, where the wheel is fixed. The adjustment for the focus is now only necessary; which is obtained by pushing downwards or upwards the proper tube, till the object appear quite distinct. In viewing transparent objects, the instrument may be used in two positions; one vertical, when the light is to be reflected upon the object by the mirror; the other, by looking up directly against the light of a candle, common light, &c.; in which case the mirror must be taken away. In viewing opaque objects, the mirror is not used; but as much common light as possible must be admitted through the circular holes in the sides of the tubes.

There is a spare hole in the transparent wheel, and also one in the opaque, to receive any occasional object that is to be viewed. Any sort of object whatsoever may be viewed, by only pushing up the microscope tube into its exterior, and bringing the first eye-tube to its focal distance from the object.

The brass tubes are so contrived, that they stop when drawn out to the full length: so that by applying one hand to the outside tube, and the other to the end of the smallest tube, the telescope at one pull may be drawn out; then any of the tubes (that next to the eye is best) may be pushed in gradually, till the most distinct view of the object be obtained.

The tubes all slide through short brass spring tubes, any of which may be unscrewed from the ends of the sliding tubes by means of the milled edges which project above the tubes, taken from each other, and the springs be clear if required.

III. *Of Solar Microscopes.*

This instrument, in its principle, is composed of a tube, a looking-glass or mirror, a convex lens, and Wilson's single microscope before described. The sun's rays being reflected through the tube by means of the mirror upon the object, the image or picture of the object is thrown distinctly and beautifully upon a screen of white paper or a white horn sheet, placed at

Plate CCC.

(A) Notwithstanding the properties that have been ascribed to the above instrument, and the praises bestowed upon it by some, which induced us to admit fo minute a description; we must apprise our readers, that it has been omitted in Mr Adams's enumeration: and upon inquiry we learn, that it has fallen into neglect among the most judicious opticians, being found too imperfect to serve the purposes of science, and too complicated for the use of persons who seek only entertainment.

Microscope it out thereon either with a pen or pencil, as it appears before them. It is worth the while of those who are desirous of taking many draughts in this way, to get a frame, wherein a sheet of paper may be put in or taken out at pleasure; for if the paper be single, the image of an object will be seen almost as plainly on the back as on the fore-side; and, by standing behind the screen, the shade of the hand will not obstruct the light in drawing, as it must in some degree when one stands before it." This construction, however, has now become rather obsolete, and is superseded by the following.

II. *The improved Solar Microscope, as used with the improved single Microscope, with teeth and pinion.* Fig. 22. represents the whole form of the *single* microscope; the parts of which are as follows: ABCD the external tube; GHIK the internal moveable one; QM part of another tube within the last, at one end of which is fixed a plate of brass hollowed in the middle, for receiving the glass tubes: there is also a moveable flat plate, between which, and the fixed end of the second tube, the ivory sliders are to be placed. L, a part of the microscope, containing a wire spiral spring, keeping the tube QM with its plates firm against the fixed part IK of the second tube.

EF is the small rack-work of teeth and pinion, by which the tube IG is moved gradually to or from the end AB, for adjusting the objects exactly to the focus of different lengths. NO is a brass slider, with six magnifiers; any one of which may easily be placed before the object. It is known when either of the glasses is in the centre of the eye-hole, by a small spring falling into a notch in the side of the slider, made against each of the glasses. These parts of the apparatus, fig. 14. (Plate xcix.) marked n° 13, 16, 17, 18, 19, 20, 21, and 22, are made use of here to this microscope. GH is a lens cell, which holds an illuminating glass for converging the sun's beams or the light of a candle strongly upon the objects. The aperture of the glass is made greater or less, by two circular pieces of brass, with holes of different sizes, that are screwed separately over the said lens. But at times, objects appear best when the microscope is held up to the common light only, without this glass. It is also taken away when the microscope is applied to the apparatus now to be described.

Fig. 23. represents the apparatus, with the single microscope screwed to it, which constitutes the *Solar Microscope.* AB is the inner moveable tube, to which the single microscope is screwed. CD, is the external tube, containing a condensing convex glass at the end D, and is screwed into the plate EF, which is cut with teeth at its circumference, and moved by the pinion I, that is fixed with the plate GH. This plate is screwed fast against the window-shutter, or board fitted to a convenient window of a darkened room, when the instrument is used. KL is a long frame, fixed to the circular plate EF; containing a looking-glass or mirror for reflecting the solar rays through the lens in the body of the tube D. O is a brass milled head, fastened to a worm or endless screw; which on the outside turns a small wheel, by which the reflecting mirror M is moved upwards or downwards.

In using this microscope, the square frame GH is first to be screwed to the window-shutter, and the
N° 218.

room well darkened: which is best done by cutting Microscope a round hole of the size of the moveable plate EF, that carries the reflector, in the window-shutter or board; and, by means of two brass nuts a a, let into the shutter to receive the screws PP, when placed through the holes in the square frame GH, at the two holes QQ; which will firmly fasten the microscope to the shutter, and is easily taken away by only unscrewing the screws PP.

The white paper screen, or white cloth, to receive the images, is to be placed several feet distant from the window: which will make the representations the larger in proportion to the distance. The usual distances are from 6 to 16 feet.

The frame KL, with its mirror M, is to be moved by turning the pinion I, one way or the other, till the beams of the sun's light come through the hole into the room: then, by turning off the worm at O, the mirror must be raised or depressed till the rays become perfectly horizontal, and go straight across the room to the screen. The tube CD, with its lens at D, is now to be screwed into the hole of the circular plate EF: by this glass the rays will be converged to a focus; and from thence proceed diverging to the screen, and there make a large circle of light. The single microscope, fig. 22. is to be screwed on to the end AB (fig. 23.) of the inner tube; and the slider NO, with either of the lenses marked 1, 2, 3, 4, 5, or 6, in the centre of the hole at the end AB. This will occasion a circle of light upon the screen much larger than before. The slider or glass-tube, with the objects to be viewed, is to be placed between the plates at IK against the small magnifier, and moved at pleasure. By shifting the tube AB in or out, you may place the object in such a part of the condensed rays as shall be sufficient to illuminate it, and not scorch or burn it; which will generally require the glass to be about one inch distant from the focus. It now remains only to adjust the object, or to bring it so near to the magnifier that its image formed upon the screen shall be the most distinct or perfect: and it is effected by gently turning the pinion F, fig. 22, a small matter one way or the other. If the object be rather large in size, the least magnifiers are generally used, and vice versâ.

N° 1. is the greatest magnifier, and n° 6. the least, in the brass slider NO. But, if desired, single lenses of greater magnifying powers are made: and they are applied, by being screwed to the end AB, fig. 22. and the brass slider NO is then taken away.

The same object may be variously magnified, by the lenses severally applied to it; and the degree of magnifying power is easily known by this rule: *As the distance of the object is to that of its image from the magnifier; so is the length or breadth of the object to that of the image.*

Instead of the brass sliders with the lenses NO, there is sometimes screwed a lens of a large size, and longer focal distance: the instrument is then converted into a *megascope;* and is adapted for viewing the larger kind of objects contained in large sliders, such as is represented at R. And, in the same manner, small objects of entertainment, painted upon glass like the sliders of a magic lanthorn, are much magnified, and represented upon the same screen.

The solar microscopes just described are capable, only

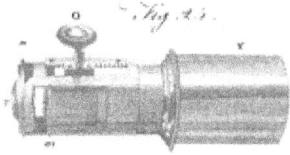

Fig. 25.

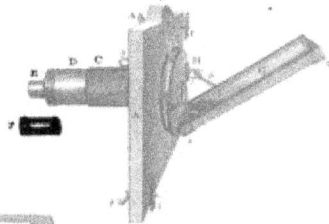

Fig. 21.

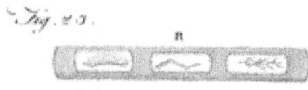

Fig. 23.

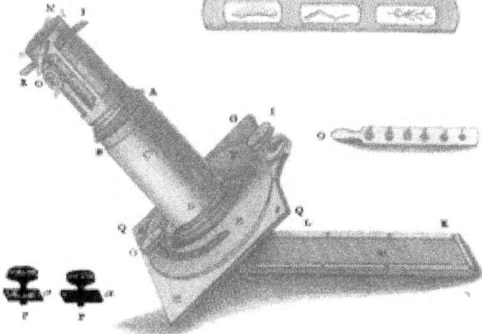

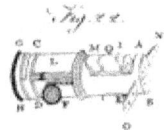

Fig. 22.

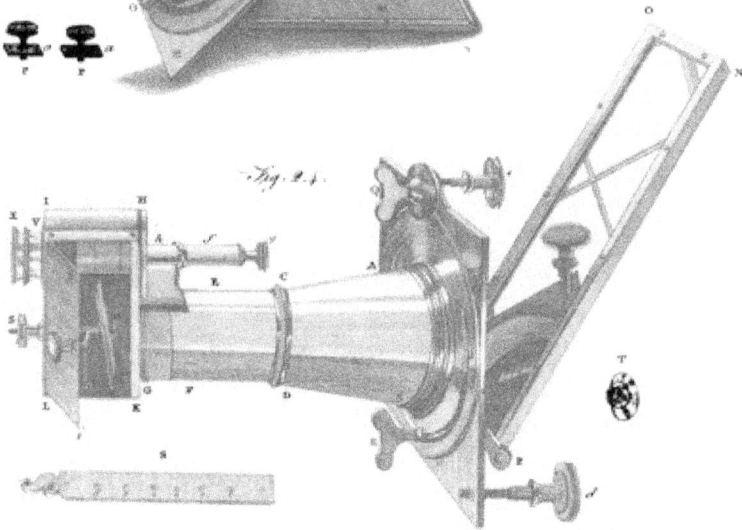

Fig. 24.

A. Bell Edin. H.S. sculpsit. fecit.

sibility of magnifying transparent objects; for which purpose the last instrument is extremely well adapted.

But as opaque objects form the most considerable part of the curious collections in the works of art as well as nature, a solar microscope for this purpose was a long time wanted.—For several years previous to 1774, the late Mr Martin made several essays towards the construction of such an instrument; and at last completed one about the time just mentioned, which he named,

III. *The Opaque Solar Microscope.* With this instrument (to use his own words) all *opaque objects*, whether of the animal, vegetable, or mineral kingdom, may be exhibited in great perfection, in all their native beauty; the lights and shades, the prominences and cavities, and all the varieties of different hues, tints, and colours; heightened by reflection of the solar rays condensed upon them."—*Transparent objects* are also shown with greater perfection than by the common solar microscope.

Fig. 24. represents the solar opaque microscope, mounted for exhibiting opaque objects.

Fig. 25. is the single tooth-and-pinion microscope, as before, which is used for showing transparent objects; the cylindrical tube Y thereof being made to fit into the tube FE of the solar microscope.

ABCDEF, (fig. 24.) represents the body of the solar microscope: one part thereof, ABCD, is conical; the other, CDEF, is cylindrical. The cylindrical part receives the tube G of the opaque box, or the tube Y of the single microscope. At the large end AB of the conical part, there is a lens to receive the rays from the mirror, and refract them towards the box HIKL. NOP is a brass frame; which is fixed to the moveable circular plate *abc*: in this frame there is a plane mirror, to reflect the solar rays on the aforementioned lens. This mirror may be moved into the most convenient position for reflecting the light, by means of the nuts Q and R. By the nut Q it may be moved from east to west; and it may be elevated or depressed by the nut R. *de*, Two screws to fasten the microscope to a window-shutter. The box for opaque objects is represented at HIKL; it contains a plane mirror M, for reflecting the light which it receives from the large lens to the object, and thereby illuminating it; S is a screw to adjust this mirror, or place it at a proper angle for reflecting the light. VX, two tubes of brass, one sliding within the other, the exterior one in the box HIKL; these carry the magnifying lenses; the interior tube is sometimes taken out, and the exterior one is then used by itself. Part of this tube may be seen in the plate within the box HIKL. At H there is a brass plate, the back part of which is fixed to the hollow tube *b*, in which there is a spiral wire, which keeps the plate always bearing against the side H of the brass box HIKL. The sliders, with the opaque objects, pass between this plate and the side of the box; to put them there, the plate is to be drawn back by means of the nut *g*: *i* is a door to one side of the opaque box. The foregoing pieces constitute the several parts necessary for viewing opaque objects. We shall now proceed to describe the single microscope, which is used for transparent objects: but in order to examine these, the box HIKL must be first removed,

VOL. XI. Part II.

and in its place we must insert the tube Y of the single microscope that we are now going to describe.

Fig. 25. represents a large tooth-and-pinion microscope; at *m*, within the body of this microscope, are two thin plates, that are to be separated, in order to let the ivory sliders pass between them; they are pressed together by a spiral spring, which bears up the under plate, and forces it against the upper one.

The slider S (under fig. 25.) which contains the magnifiers, fits into the hole *n c* and any of the magnifiers may be placed before the object, by moving the aforesaid slider: when the magnifier is at the centre of the hole P, a small spring falls into one of the notches which is on the side of the slider.

Under the plate *n* are placed two lenses, for enlarging the field of view on the screen; the smaller of the two is fixed at a piece of brass, and is moved the plate *o*; this is to be taken out when the magnifiers, N° 4, 5, or 6, are used, or when the megalascope lens T (fig. 24.) is used; but it is to be replaced in N° 1, 2, 3.

This microscope is adjusted to the focus by turning the milled nut O.

To use the solar microscope:—Make a round hole in the window-shutter, a little larger than the circle *abc*; pass the mirror ONP through this hole, and apply the square plate to the shutter; then mark with a pencil the places which correspond to the two holes through which the screw is to pass; take away the microscope, and bore two holes at the marked places, sufficiently large to let the milled screws *de* pass through them.

The screws are to pass from the outside of the shutter, to go through it; and being then screwed into their respective holes in the square plate, they will, when screwed home, hold it fast against the inside of the shutter, and thus support the microscope.

Screw the conical tube ABCD to the circle *abc*, and then slide the tube G of the opaque box into the cylindrical part CD EF of the body, if opaque objects are to be examined; but if they be transparent objects you mean to show, then place the tube Y within the tube CDEF.

The room is to be darkened as much as possible, that no light may enter but what passes through the body of the microscope; for, on this circumstance, together with the brightness of the sun shine, the perfection and distinctness of the image in a great measure depend.

When the microscope is to be used for opaque objects, 1. Adjust the mirror NOP, so as to receive the solar rays, by means of the two finger screws or nuts, Q R, the first, Q, turns the mirror to the right or left; the second, R, raises or depresses it: this you are to do till you have reflected the sun's light through the lens at AB strongly upon a screen of white paper placed at some distance from the window, and formed thereon a round spot of light. As unexperienced observer will find it more convenient to obtain the light by forming this spot before he puts on either the opaque box or the tooth-and-pinion microscope.

Now put in the opaque box, and place the object between the plates at H; upon the door *ik*, and adjust the mirror M till you have illuminated the object strongly. If you cannot effect this by the screw S,

4 Y   you

Microscope you must move the screws Q, R, in order to get the light reflected strongly from the mirror NOP, or the mirror M, without which the latter cannot illuminate the object.

The object being strongly illuminated, that the door *b, and a distinct view of the object will soon be obtained on your screen, by adjusting the tubes VX, which is effected by moving them backwards or forwards.

A round spot of light cannot always be procured in northern latitudes, the altitude of the sun being often too low; neither can it be obtained when the sun is directly perpendicular to the front of the room.

As the sun is continually changing its place, it will be necessary, in order to keep his rays full upon the object, to keep them continually directed thro' the axis of the instrument, by the two screws Q and R.

To view transparent objects, remove the opaque box, and insert the tube Y, fig. 25. in its place; put the slider S into its place at *n, and the slider with the objects between the plates at *m; then adjust the mirror NOP, as before directed by the screws Q, R, so that the light may pass through the object; regulate the focus of the magnifier by the screw O. The most pleasing magnifiers in use are the fourth and fifth.

The size of the object may be increased or diminished, by altering the distance of the screen from the microscope; five or six feet is a convenient distance.

To examine transparent objects of a larger size, or to render the instrument what is usually called a megalascope, take out the slider b from its place at *n, and screw the button T (fig. 24.) into the hole at P, fig. 25. and remove the glass which is under the plate at *m, and regulate the light and focus agreeable to the foregoing directions.

N. B. At the end of the tube G there is a lens for increasing the density of the rays, for the purpose of burning or melting any combustible or fusible substance; this lens must be removed in most cases, lest the objects should be burnt. The intensity of the light is also varied by moving this tube backwards or forwards.

Apparatus of the Opaque Solar Microscope.—The large square plate and mirror; the body of the microscope; the opaque box and its tube; the tooth-and-pinion microscope; the slider with the magnifiers; the megalascope magnifier; the two screws Q and R; some ivory sliders; bone sliders with opaque objects; a brass frame, with a bottom of soft deal to stick any object on; a brass cylinder X (fig. 31.), for confining opaque objects.

IV. *The Camera Obscura, or Lucernal, Microscope.*

—The great facility with which objects can be represented on paper or a rough glass in the camera obscura, and copies drawn from them by any person though unskilled in drawing, evidently suggested the application of the microscope to this instrument. The greatest number of experiments that appear to have been made with this view, were by the late Mr Martin and Mr Adams; the former of whom frequently applied the microscope to the portable camera, and with much effect and entertainment. But these instruments being found to answer only with the assistance of the sun, Mr Adams directed his experiments to the construc-

tion of an instrument of more extended utility, which could be equally employed in the day-time and by night. He accordingly succeeded so far as to produce, by candle-light, the images of objects refracted from a single magnifier upon one or two large convex lenses (of about five inches or upwards in diameter), at the end of a pyramidal shaped box, in a very pleasing and magnified appearance, so as to give opaque objects as well as transparent ones the utmost distinctness of representation; but still the light of a candle or lamp was found generally insufficient to throw the requisite degree of illumination upon the objects. The invention of what is called *Argand's lamp*, within these few years offered a complete remedy for this defect, by the intensity and steadiness of its light. This did not escape the present Mr Adams (son of the former), who immediately applied it; and who had likewise so altered and improved his father's instrument, both in construction and form, as to render it altogether a different one, and far more perfect and useful.

The advantages and properties of this excellently conceived instrument are numerous and important. "As the far greater part of the objects which surround us are opaque (says our author), and very few are sufficiently transparent to be examined by the common microscope, an instrument that could be readily applied to the examination of opaque objects has always been a desideratum. Even in the examination of transparent objects, many of the fine and more curious portions are lost, and drowned, as it were, in the light which must be transmitted through them; while different parts of the same object appear only as dark lines or spots, because they are so opaque as not to permit any light to pass through them. These difficulties, as well as many more, are obviated in the lucernal microscope; by which opaque objects of various sizes may be seen with ease and distinctness: the beautiful colours with which most of them are adorned, are rendered more brilliant, without changing in the least the real tint of the colour; and the concave and convex parts retain also their proper form.—The facility with which all opaque objects are applied to this instrument, is another considerable advantage, and almost peculiar to itself; as the texture and configuration of the more tender parts are often hurt by previous preparation, every object may be examined by this instrument, first as opaque, and afterwards (if the texture will admit of it) as transparent.—The lucernal microscope does not in the least fatigue that eye; the object appears like nature itself, giving ease to the sight and pleasure to the mind: there is also, in the use of this instrument, no occasion to shut the eye which is not directed to the object. A further advantage peculiar to this microscope is, that by it the outlines of every object may be taken, even by those who are not accustomed to draw; while those who can draw well will receive great assistance, and execute their work with more accuracy and in less time than they would otherwise have been able to have performed it. Transparent objects as well as opaque may be copied in the same manner. The instrument may be used at any time of the day, but the best effect is by night; in which respect it has a superiority over the solar microscope, as that instrument can only be used when the sun shines.

3   TRAN-

Microscope. Transparent objects may be examined with the lucernal microscope in three or four different modes, from a blaze of light almost too great for the eye to bear, to that which is perfectly easy to it: And by the addition of a tin lanthorn to the apparatus, may be thrown on a screen, and exhibited at one view to a large company, as by the solar microscope.

We shall now proceed to the description of the instrument and apparatus as given by Mr Adams.

Fig. 26. represents the improved *Lucernal Microscope*, mounted to view opaque objects. A B C D is a large mahogany pyramidal box, which forms the body of the microscope; it is supported firmly on the brass pillar FG, by means of the socket H and the curved piece I K.

LMN is a guide for the eye, in order to direct it in the axis of the lenses; it consists of two brass tubes, one sliding within the other, and a vertical flat piece, at the top of which is the hole for the eye. The outer tube is seen at MN, the vertical piece is represented at LM. The inner tube may be pulled out, or pushed in, to adjust it to the focus of the glasses. The vertical piece may be raised or depressed, that the hole, through which the object is to be viewed, may coincide with the centre of the field of view; it is fixed by a milled screw at M, which could not be shewn in this figure.

At N is a dove-tailed piece of brass, made to receive the dove-tail at the end of the tubes MN, by which it is affixed to the wooden box ABCDE. The tubes MN may be removed from this box occasionally, for the convenience of packing it up in a less compass.

O P, a small tube which carries the magnifiers.

O, one of the magnifiers; it is screwed into the end of a tube, which slides within the tube P, the tube P may be unscrewed occasionally from the wooden body.

QRSTVX, a long squarebar, which passes through the sockets YZ, and carries the stage or frame that holds the objects; this bar may be moved backward or forward, in order to adjust it to the focus by means of the pinion which is at *a*.

b, A handle furnished with an universal joint, for more conveniently turning the pinion. When the handle is removed, the nut (fig. 27.) may be used in its stead.

de, A brass bar, to support the curved piece KI, and keep the body AB firm and steady.

fghi, The stage for opaque objects: it fits upon the bar QRST by means of the socket *hi*, and is brought nearer to or removed farther from the magnifying lens by turning the pinion *a*: the objects are placed in the front side of the stage (which cannot be seen in this figure) between four small brass plates; the edges of two of these are seen at *kk*. The two upper pieces of brass are moveable; they are fixed to a plate, which is acted on by a spiral spring, that presses them down, and confines the slider with the objects: this plate, and the two upper pieces of brass, are lifted up by the small nut *m*.

At the lower part of the stage, there is a semicircular lump of glass *n*, which is designed to receive the light from the lamp, fig. 29. and to collect and throw it on the concave mirror *o*, whence it is to be reflected on the object.

The upper part *fgrs* (fig. 26.) of the opaque stage takes out, that the stage for transparent objects may be inserted in its place.

Fig. 28. represents the stage for transparent objects; the two legs 5 and 6 fit into the top of the under part *rsbi* of the stage for opaque objects; 7 is the part which confines or holds the sliders, and through which they are to be moved; 9 and 10 a brass tube, which contains the lenses for condensing the light, and throwing it upon the object; there is a second tube within that, marked 9 and 10, which may be placed at different distances from the object by the pin 11.

When this stage is used as a single microscope, without any reference to the lucernal, the magnifiers, or object lenses, are to be screwed into the hole 12, and to be adjusted to a proper focus by the nut 13.

N. B. At the end AB (fig 26.) of the wooden body there is a slider, which is represented as partly drawn out at A: when quite taken out, three grooves will be perceived; one of which contains a board that forms the end of the box; the next contains a frame with a greyed glass; and the third, or that farthest from the end AB, two large convex lenses.

Fig. 29. represents one of Argand's lamps, which are the most suitable for microscopic purposes, on account of the clearness, the intensity, and the steadiness of the light. The following account of the method of managing them, with other observations, is copied from an account given by Mr Parker with those he sells.

The principle on which the lamp acts, consists in disposing the wick in thin parts, so that the air may come into contact with all the burning fuel; by which means, together with an increase of the current of air occasioned by rarefaction in the glass tube, the whole of the fuel is converted into flame.

The wicks are circular; and, the more readily to regulate the quantity of light, are fixed on a brass collar, with a wire handle, by means of which they are raised or depressed at pleasure.

To fix the wick on, a wooden mandril is contrived, which is tapered at one end, and has a groove turned at the other.

The wick has a selvage at one end, which is to be put foremost on the mandril, and moved up to the groove; then putting the groove into the collar of the wick-holder, the wick is easily pushed forward upon it.

The wick-holder and wick being put quite down in their place, the spare part of the wick should, while dry, be set a-light, and suffered to burn to the edge of the tubes; this will leave it more even than by cutting; and, being black by burning, will be much easier lighted: for this reason, the black should never be quite cut off.

The lamp should be filled an hour or two before it is wanted, that the cotton may imbibe the oil and draw the better.

The lamps which have a reservoir and valve, need no other direction for filling than to do it with a proper trimming pot, carefully observing when they are full; then pulling up the valve by the point, the reservoir, being turned with the other hand, may be replaced without spilling a drop.

Those lamps which fill in the front like a bird fountain, must be reclined on the back to fill; and this

should

Microscope should be done gently, that the oil in the burner may return into the body when so placed and filled: if, by being too full, any oil appears above the guard, only move the lamp a little, and the oil will disappear; the lamp may then be placed erect, and the oil will flow to its proper level.

The oil must be of the spermaceti kind, commonly called chamber-oil, which may generally be distinguished by its paleness, transparency, and inoffensive scent: all those oils which are of a red and brown colour, and of an offensive scent, should be carefully avoided, as their glutinous parts clog the lamp, and the impurities in such oil, not being inflammable, will accumulate and remain in the form of a crust on the wick. Seal oil is nearly as pale and sweet as chamber oil; but being of a heavy sluggish quality, is not proper for lamps with fine wicks.

Whenever bad oil has been used, on changing it, the wick must also be changed; because, after having imbibed the coarse particles in its capillary tubes, it will not draw up the fine oil.

To obtain the greatest degree of light, the wick should be trimmed exactly even, the flame will then be completely equal.

There will be a great advantage in keeping the lamp clean, especially the burner and air-tubes; the neglect of cleanliness in lamps is too common; a candlestick is generally cleaned every time it is used, so should a lamp; and if a candlestick is not to be objected to because it does not give light after the candle is exhausted, so a lamp should not be thought ill of, if it does not give light when it wants oil or cotton; but this ill has often happened, because the deficiency is less visible.

The glass tubes are best cleaned with a piece of wash leather.

If a fountain-lamp is left partly filled with oil, it may be liable to overflow; this happens by the contraction of the air when cold, and its expansion by the warmth of a room, the rays of the sun, or the heat of the lamp when re-lighted: this accident may be effectually prevented by keeping the reservoir filled, the oil not being subject to expansion like air. On this account, those with a common reservoir are best adapted for microscopic purposes.

To examine Opaque Objects with the Lucernal Microscope. To render the use of this instrument easy, it is usually packed with as many of the parts together as possible; it occupies on this account rather more room, but is much less embarrassing to the observer, who has only three parts to put on after it is taken out of its box, namely, the guide for the eye, the stage, and the tube with its magnifier.

But to be more particular: Take out the wooden slider A (fig. 26.), then lift out the cover and the grey glass from their respective grooves under the slider A.

Put the end N of the guide for the eye LMM into its place, so that it may stand in the position which is represented in this figure.

Place the socket which is at the bottom of the opaque stage, on the bar Q X T, so that the concave mirror s may be next the end DE of the wooden body.

Screw the tubes PO into the end DE. The magnifier you intend to use is to be screwed on the end O of these tubes.

The handle GS, or the milled nut fig. 27. must be placed on the square end of the pinion s.

Place the lamp lighted before the glass lamp s, and the object you intend to examine between the spring-plates of the stage; and the instrument is ready for use.

In all microscopes there are two circumstances which must be particularly attended to: first, the modification of the light, or the proper quantity to illuminate the object; secondly, the adjustment of the instrument to the focus of the glasses and eye of the observer. In the use of the lucernal microscope there is a third circumstance, which is, the regulation of the guide for the eye.

1. To throw the light upon the object. The flame of the lamp is to be placed rather below the centre of the glass lamp s, and as near it as possible; the concave mirror s must be so inclined and turned as to receive the light from the glass lamp, and reflect it thence upon the object; the best situation of the concave mirror and the flame of the lamp depends on a combination of circumstances, which a little practice will discover.

2. To regulate the guide for the eye, or to place the centre of the eye-piece L so that it may coincide with the focal point of the lenses and the axis of vision: Lengthen and shorten the tubes MN, by drawing out or pushing in the inner tube, and raising or depressing the eye-piece ML, till you find the large lens (which is placed at the end AB of the wooden body) filled by an uniform field of light, without any prismatic colours round the edge; for till this piece is properly fixed, the circle of light will be very small, and only occupy a part of the lens: the eye must be kept at the centre of the eye-piece L, during the whole of the operation; which may be rendered somewhat easier to the observer, on the first use of the instrument, if he hold a piece of white paper parallel to the large lens, removing it from or bringing it nearer to them till he find the place where a lucid circle, which he will perceive on the paper, is brightest and most distinct; then he is to fix the centre of the eye-piece to coincide with that spot; after which a very small adjustment will set it perfectly right.

3. To adjust the lenses to their focal distance. This is effected by turning the pinion s, the eye being at the same time at the eye-piece L. The grey glass is often placed before the large lenses, while regulating the guide for the eye, and adjusting for the focal distance.

If the observer, in the process of his examination of an object, advance rapidly from a shallow to a deep magnifier, he will save himself some labour by pulling out the internal tube at O.

The upper part fg r s of the stage is to be raised or lowered occasionally, in order to make the centre of the object coincide with the centre of the lens at O.

To delineate objects, the grey glass must be placed before the large lenses; the picture of the object will be formed on this glass, and the outline may be accurately taken by going over the picture with a pencil.

The

Plate CCCI.

Fig. 26.

Fig. 29.

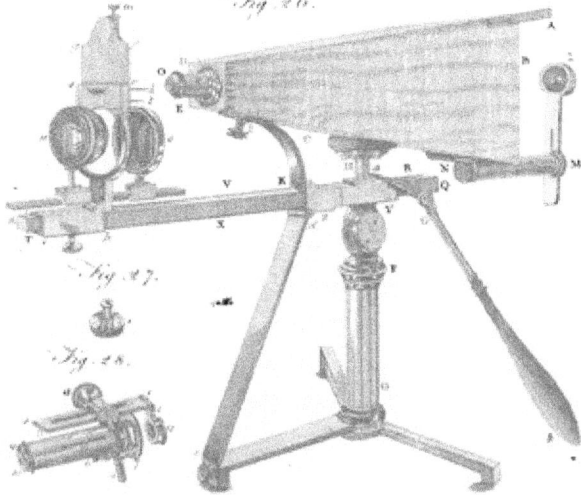

Fig. 27.

Fig. 28.

Fig. 30.

Fig. 31.

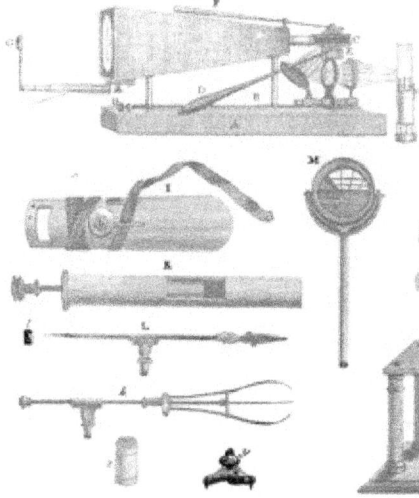

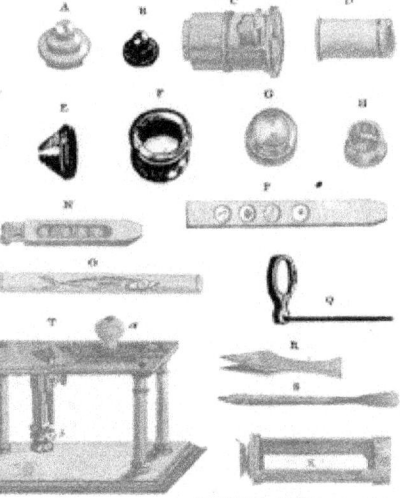

The opaque part may be used in the day-time without a lamp, provided the large lenfes at A B are fcreened from the light.

To ufe the Lucernal Microfcope in the examination of Tranfparent Objects. The inftrument is to remain as before : the upper part *f g* of the opaque ftage muft be removed, and the ftage for tranfparent objects, reprefented at fig. 28. put in its place ; the end 9 10 to be next the lamp.

Place the greyed glafs in its groove at the end A B, and the objects in the flider-holder at the front of the ftage ; then tranfmit as ftrong a light as you are able on the object, which you will eafily do by raifing or lowering the lamp.

The object will be beautifully depicted on the grey glafs : it muft be regulated to the focus of the magnifier, by turning the pinion *o*.

The object may be viewed either with or without the guide for the eye. A fingle obferver will fee an object to the greateft advantage by ufing this guide, which is to be adjufted as we have defcribed above. If two or three wifh to examine the object at the fame time, the guide for the eye muft be laid afide.

Take the large lens out of the groove, and receive the image on the grey glafs ; in this cafe, the guide for the eye is of no ufe : if the grey glafs be taken away, the image of the object may be received on a paper fcreen.

Take out the grey glafs, replace the large lenfes, and ufe the guide for the eye ; attend to the foregoing directions, and adjuft the object to its proper focus. You will then fee the object in a blaze of light almoft too great for the eye, a circumftance that will be found very ufeful in the examination of particular objects. The edges of the object in this mode will be fomewhat coloured : but as it is only ufed in this full light for occafional purpofes, it has been thought better to leave this fmall imperfection, than, by remedying it, to facrifice greater advantages ; the more fo, as this fault is eafily corrected, and a new and interefting view of the object is obtained, by turning the inftrument out of the direct rays of light, and permitting them to pafs through only in an oblique direction, by which the upper furface is in fome degree illuminated, and the object is feen partly as opaque, partly as tranfparent. It has been already obferved, that the tranfparent objects might be placed between the flider-holders of the ftage for opaque objects, and then be examined as if opaque.

Some tranfparent objects appear to the greateft advantage when the lens at 9 10 is taken away ; as, by giving too great a quantity of light, it renders the edges lefs fharp.

The variety of views which may be taken of every object by means of the improved lucernal microfcope, will be found to be of great ufe to an accurate obferver : it will give him an opportunity of correcting or

confirming his difcoveries, and inveftigating thofe parts in one mode which are invifible in another.

To throw the image of tranfparent objects on a fcreen, as in the folar microfcope. It has been long a microfcopical defideratum, to have an inftrument by which the image of tranfparent objects might be thrown on a fcreen, as in the common folar microfcope : and this not only becaufe the fun is fo uncertain in this climate, and the ufe of the folar microfcope requires confinement in the fineft part of the day, when time feldom hangs heavy on the mind ; but as it alfo affords an increafe of pleafure, by difplaying its wonders to feveral perfons at the fame inftant, without the leaft fatigue to the eye.

This purpofe is now effectually anfwered, by affixing the tranfparent ftage of the lucernal to a lanthorn, with one of Argand's lamps.—The lamp is placed within the lanthorn, and the end 9 10 of the tranfparent ftage is fcrewed into a female fcrew, which is rivetted in the fliding part of the front of the lanthorn ; the magnifying lenfes are to be fcrewed into the hole reprefented at 12, and they are adjufted by turning the milled nut. The quantity of light is to be regulated by raifing and lowering the fliding-plate or the lamp.

Apparatus which ufually accompanies the improved Lucernal Microfcope. The ftage for opaque objects, with its femicircular lamp of glafs, and concave mirror. The ftage for tranfparent objects, which fits on the upper part of the foregoing ftage. The fliding tube, to which the magnifiers are to be affixed : one end of thefe is to be fcrewed on the end D of the wooden body ; the magnifier in ufe is to be fcrewed to the other end of the inner tube. Eight magnifying lenfes : thefe are fo conftructed, that they may be combined together, and thus produce a very great variety of magnifying powers. A fifh-pan, fuch as is reprefented at I. A fteel wire L, with a pair of nippers at one end, and a fmall cylinder of ivory I at the other. A flider of brafs N, containing a flat glafs flider, and a brafs flider into which are fitted fome fmall concave glaffes. A pair of forceps. Six large and fix fmall ivory fliders, with tranfparent objects. Fourteen wooden fliders, with four opaque objects in each flider ; and two fpare fliders. Some capillary tubes for viewing fmall animalcula.

Ingenious men feldom content themfelves with an inftrument under one form ; hence fuch a variety of microfcopes, hence many alterations in the Lucernal Microfcope. Mr Adams himfelf, we underftand, has fitted up this laft in a great many different ways ; and it is reafonable to think that no perfon is more likely to give it every improvement of which it is fufceptible. Of the alterations by other hands we fhall only particularife one, made by Mr Jones of Holborn (a), whofe defcription is as follows :

A, reprefents a portion of the top of the mahogany box

(a) We truft the reader will never confider any paragraph wherein the name of an inftrument-maker or other artift is inferted, as a recommendation of thofe artifts by the editors of this work. In the courfe of a pretty extenfive correfpondence, they have been favoured with very liberal communications from various artifts, for which they are greatly indebted to them : the inferting their names in this work is therefore to be confidered as a grateful acknowledgment from the editors for favours conferred on them,—not as a teftimonial

Microscope box in which it packs, to preserve it steady; it slides in a dove-tail groove withinside, a similar groove to which is cut in the top of the box A; so that when the instrument is to be used, it is slipt out of the box withinside, and then slipt into the groove at top ready for use, almost instantly, as shown in the figure. The adjustment of the objects is at the stage E; for the right focal distance is readily and conveniently made by turning the long screw-rod B B, which goes thro' the two pillars supporting the box, and works in the base of the brass stage E; which base is also dove-tailed, so as to have a regular and steady motion in another brass basis that supports it. In this instrument, therefore, the pyramidical box does not move; but the stage part only, which, from its small weight, moves in the most agreeable and steady manner. While observing the image of the object upon the glass through the sight-hole at G, the object may be moved or changed by only turning the rack-work and pinion applied to the stage, by means of the handle D, for that purpose. By this contrivance you have no occasion to change your position during the view of the objects upon one of the sliders. This motion changes the objects horizontally only; and as they are generally placed exactly in one line, it answers all the purposes for which this motion is intended very well. But it may sometimes happen that the observer would wish to alter the vertical position of the object; to perform which there is another plane rod at F, that acts simply as a lever for this purpose, and moves the sliding part of the stage E vertically either upwards or downwards.

Thus, without altering his position, the observer may investigate all parts of the objects in the most satisfactory manner. Rack-work and pinion might be applied to the stage for the vertical motion also; but as it would materially enhance the expence, it is seldom applied. The brass work at the handle of D contains a Hooke's universal joint.

The brilliancy of the images of the objects shown upon the large lenses at the end of the box, being very frequently so great as to dazzle the eyes, Mr Jones applies a slight tinge of blue, green, and other coloured glass, to the sight-hole at G, which softens this glare, and casts an agreeable hue upon the objects.

Description of those Parts of a Microscopical Apparatus, common to most Instruments, which are delineated at fig. 31.

A and B represent the brass cells which contain the magnifiers belonging to the different kinds of compound microscopes. The magnifiers are sometimes contained in a slider like that which is delineated at S (fig. 24). The lenses of A and B are confined by a small cap; on unscrewing this, the small lens may be taken out and cleaned. The magnifiers A of the lucernal microscope are so contrived, that any two of them may be screwed together, by which means a considerable variety of magnifying power is obtained.

To get at the lenses in the slider S (fig. 24.), take out the two screws which hold on the cover.

C, represents the general form of the slider-holder. It consists of a cylindrical tube, in which an inner tube is forced up by a spring. It is used to receive the ivory or any other slider, in which the transparent objects are placed; these are to be slid between the two upper plates: the hollow part in one of the plates is designed for the glass tubes.

D, the condensing lens and its tube, which fits into the slider-holder C, and may be moved up and down in it. When this piece is pushed up as far as it will go, it condenses the light of a candle, which is reflected on it by the plain mirror of the compound microscope, and spreads it uniformly over the object; in this case it is best adapted to the shallowest magnifiers. If the deeper lenses are used, it should be drawn down, or rather removed farther from the object, that it may concentrate the light in a small compass, and thus render it more dense. The condensing lens is sometimes fitted up differently; but the principle being the same, it will be easy to apply it to use notwithstanding some variations in the mechanism.

E, a brass cone. It fixes under the slider-holder, and is used to lessen occasionally the quantity of light which comes from the mirror to any object.

F, a box with two flat glasses, which may be placed at different distances from each other in order to confine a small living insect.

G, a small brass box to hold the silver speculum H.

H, a small silver concave speculum, designed to reflect the light from the mirror on opaque objects; it should only be used with the shallow magnifiers. It is applied in different ways to the compound microscope; sometimes to a tube similar to that represented at X, which slides on the lower part of the body; sometimes it is screwed into the ring of the piece Q; the pin of this generally fits into one of the holes in the stage. When this speculum is used, the slider-holder should be removed.

I, a fish-pan, whereon a small fish may be fastened, in order to view the circulation of the blood; its tail is to be spread across the oblong hole at the smallest end, and tied fast by means of the ribbon fixed thereto, by shoving the knob which is on the back of it through the slit made in the stage; the tail of the fish may be brought under the lens which is in use.

K, a cylindrical piece, intended for the solar opaque microscope; by pulling back the spiral spring, smaller or larger objects may be confined in it.

k, A pair of triangular nippers for taking hold of and confining a large object.

L, a long steel wire, with a small pair of pliers at one end and a steel point at the other: the wire slips backwards or forwards in a spring tube, which is affixed to a joint, at the bottom of which is a pin to fit one of the holes in the stage; this piece is used to confine small objects.

l, A small ivory cylinder that fits on the pointed end of the steel wire L; it is designed to receive opaque objects. Light-coloured ones are to be stuck on the dark side, and vice versa.

M, a convex lens, which fits to the stage by means of

---

nial of their opinion of the abilities of an individual, or as designed to insinuate any preference over others in the same line, where such preference has not been already bestowed by the public.

of the long pin adhering to it. This piece is designed to collect the light from the sun or a candle, and to throw them on any object placed on the stage; but it is very little used at present.

N, a brass slider, into which is fitted a flat piece of glass, and a brass slider containing four small glasses, one or two of them concave, the others flat; it is designed to confine small living objects, and when used is to be placed between the two upper plates of the slider-holder.

O, a glass tube to receive a small fish, &c.

P, represents one of the ivory sliders, wherein objects are placed between two pieces of talc, and confined by a brass ring.

Q, a piece to hold the speculum H: this piece is generally fitted to the microscope represented at fig. 12.

R, a pair of forceps, to take up any occasional object.

S, a camel's hair pencil to brush the dust off the glasses; the upper part of the quill is scooped out, to take up a drop of any fluid, and place it on either of the glasses for examination.

T, an instrument for cutting thin transverse sections of wood. It consists of a wooden base, which supports four brass pillars; on the top of the pillars is placed a flat piece of brass, near the middle of which there is a triangular hole.

A sharp knife, which moves in a diagonal direction, is fixed on the upper side of the afore-mentioned plate, and in such a manner that the edge always coincides with the surface thereof.

The knife is moved backwards and forwards by means of the handle a. The piece of wood is placed in the triangular trough which is under the brass plate, and is to be kept steady therein by a milled screw which is fitted to the trough; the wood is to be pressed forward for cutting by the micrometer screw b.

The pieces of wood should be applied to this instrument immediately on being taken out of the ground, or else they should be soaked for some time in water, to soften them so that they may not hurt the edge of the knife.

When the edge of the knife is brought in contact with the piece of wood, a small quantity of spirits of wine should be poured on the surface of the wood, to prevent its curling up; it will also make it adhere to the knife, from which it may be removed by pressing a piece of blotting paper on it.

y, An appendage to the cutting engine, which is to be used instead of the micrometer screw, being preferred to it by some. It is placed over the triangular hole, and kept flat down upon the surface of the brass plate, while the piece of wood is pressed against a circular piece of brass which is on the under side of it. This circular piece of brass is fixed to a screw, by which its distance from the flat plate on which the knife moves may be regulated.

z, An ivory box, containing at one end spare talc for the ivory sliders, and at the other spare rings for pressing the talcs together and confining them to the slider.

AFTER what has been related of Microscopes, they cannot be said to be complete without the valuable

addition of a *micrometer*; for the use and advantages Microscope of which, see the article MICROMETER.

HAVING presented our readers with descriptions of the various microscopes generally used, we think it our duty to point out to them those which we conceive to be best calculated to answer the purposes of science. The first which presents itself to our mind is that of *Lieb*: It is better adapted, than any other portable microscope, to the purpose of general observation; simple in its construction, and general in its application. To those who prefer a double microscope, we should recommend that figured in Plate CCXCVIII. (12.) If opaque objects, as insects, &c. be subjects of investigation, the *Lucernal Microscope* claims the preference: but if amusement alone guides the choice, the *Solar Microscope* must be fixed upon.

We shall now proceed to explain some necessary particulars respecting the method of using microscopes; after which, we shall subjoin an enumeration of the principal objects discovered or elucidated by their means. On this subject Mr Adams, in his *Essay on the Microscope*, has been very copious; with a view, as he informs us, to remove the common complaint made by Mr Baker, "that many of those who purchase microscopes are so little acquainted with their general and extensive usefulness, and so much at a loss for objects to examine by them, that after diverting their friends some few times with what they find in the sliders which generally accompany the instrument, or perhaps with two or three common objects, the microscope is laid aside as of little further value; whereas no instrument has yet appeared in the world capable of affording so constant, various, and satisfactory an entertainment to the mind."

I. In using the microscope, there are three things necessary to be considered. (1.) The preparation and adjustment of the instrument itself. (2.) The proper quantity of light, and the best method of adapting it to the object. (3.) The method of preparing the objects, so that their texture may be properly understood.

1. With regard to the microscope itself, the first thing necessary to be examined is, whether the glasses be clean or not; if they are not so, they must be wiped with a piece of soft leather, taking care not to soil them afterwards with the fingers; and, in replacing them, care must be taken not to place them in an oblique situation. We must likewise be careful not to let the breath fall upon the glasses, nor to hold that part of the body of the instrument where the glasses are placed with a warm hand; because thus the moisture expelled by the heat from the metal will condense upon the glass, and prevent the object from being distinctly seen. The object should be brought as near the centre of the field of view as possible; for there only it will be exhibited in the greatest perfection. The eye should be moved up and down from the eye-glass of a compound microscope, till the situation is found where the largest field and most distinct view of the object are to be had; but every person ought to adjust the microscope to his own eye, and not to depend upon the situation it was placed in by another. A small magnifying power should

Microscope always be begun with; by which means the observer will best obtain an exact idea of the situation and connection of the whole; and will of consequence be less liable to form any erroneous opinion when the parts are viewed separately by a lens of greater power. Objects should also be examined first in their most natural positions: for if this be not attended to, we shall be apt to form very inadequate ideas of the structure of the whole, as well as of the connection and use of the parts. A living animal ought to be as little hurt or discomposed as possible.

From viewing an object properly, we may acquire a knowledge of its nature: but this cannot be done without an extensive knowledge of the subject, much patience, and many experiments; as in a great number of cases the images will resemble each other, though derived from very different substances. Mr Baker therefore advises us not to form an opinion too suddenly after viewing a microscopical object; nor to draw our inferences till after repeated experiments and examinations of the object in many different lights and positions; to pass no judgment upon things extended by force, or contracted by dryness, or in any manner out of a natural state, without making suitable allowances. The true colour of objects cannot be properly determined by very great magnifiers; for as the pores and interstices of an object are enlarged according to the magnifying power of the glasses made use of, the component particles of its substance will appear separated many thousand times farther asunder than they do to the naked eye: hence the reflection of the light from these particles will be very different, and exhibit different colours. It is likewise somewhat difficult to observe opaque objects; and as the apertures of the larger magnifiers are but small, they are not proper for the purpose. If an object be to very opaque, that no light will pass through it, as much as possible must be thrown upon the upper surface of it. Some consideration is likewise necessary in forming a judgment of the motion of living creatures, or even of fluids, when seen through the microscope; for as the moving body, and the space wherein it moves, are magnified, the motion will also be increased.

2. On the management of the light depends in a great measure the distinctness of the vision; and as, in order to have this in the greatest perfection, we must adapt the quantity of light to the nature of the object and the focus of the magnifier, it is therefore necessary to view it in various degrees of light. In some objects, it is difficult to distinguish between a prominence and a depression, a shadow or a black stain; or between a reflection of light and whiteness, which is particularly observable in the eye of the libella and other flies: all of these appearing very different at one position from what they do in another. The brightness of an object likewise depends on the quantity of light, the distinctness of vision, and on regulating the quantity to the object; for some will be in a manner lost in a quantity of light scarce sufficient to render another visible.

There are various ways in which a strong light may be thrown upon objects; as by means of the sun and a convex lens. For this purpose, the microscope is to be placed about three feet from a southern window;

N 249.

then take a deep convex lens, mounted on a semicircle and stand, so that its position may easily be varied: place this lens between the object and the window, so that it may collect a considerable number of solar rays, and reflect them on the object or the mirror of the microscope. If the light thus collected from the sun be too powerful, it may be lessened by placing a piece of oiled paper, or a piece of glass lightly greyed, between the object and lens. Thus a proper degree of light may be obtained, and diffused equally all over the surface of an object: a circumstance which ought to be particularly attended to; for if the light be thrown irregularly upon it, no distinct view can be obtained. If we mean to make use of the solar light, it will be found convenient to darken the room, and to reflect the rays of the sun on the abovementioned lens by means of the mirror of a solar microscope fixed to the window-shutter; for thus the observer will be enabled to preserve the light on his object, notwithstanding the motion of the sun. But by reason of this motion, and the variable state of the atmosphere, solar observations are rendered both tedious and inconvenient; whence it will be proper for the observer to be furnished with a large tin lanthorn, formed something like the common magic lanthorn, capable of containing one of Argand's lamps. This, however, ought not to be of the fountain kind, lest the rarefaction of the air in the lanthorn should force the oil over. There ought to be an aperture in the front of the lanthorn, which may be moved up and down, and be capable of holding a lens; by which means a pleasant and uniform as well as strong light may easily be procured. The lamp should likewise move on a rod, so that it may be easily raised or depressed. This lanthorn may likewise be used for many other purposes; as viewing of pictures, exhibiting microscopic objects on a screen, &c. A weak light, however, is best for viewing many transparent objects; among which we may reckon the prepared eyes of flies, as well as the animalcules in fluids. The quantity of light from a lamp or candle may be lessened by removing the microscope to a greater distance from them, or by diminishing the strength of the light which falls upon the objects. This may very conveniently be done by pieces of black paper with circular apertures of different sizes, and placing a larger or smaller one upon the reflecting mirror, as occasion may require. There is an oblique situation of the mirror, which makes likewise an oblique reflection of the light easily discovered by practice, (but for which no general rule can be given in theory); and which will exhibit an object more distinctly than any other position, showing the surface, as well as those parts through which the light is transmitted. The light of a lamp or candle is generally better for viewing microscopic objects than day-light; it being more easy to modify the former than the latter, and to throw it upon the objects with different degrees of density.

3. With regard to the preparation of objects, Swammerdam has, in that particular, excelled almost all other investigators who either preceded or have succeeded him. He was so assiduous and indefatigable, that neither difficulty nor disappointment could make the least impression upon him; and he never abandoned the pursuit of any object until he had obtained

Microscope be fastened by a piece of soft wax, and again covered with water.—Larger objects should be placed in a trough of thin wood; and for this purpose the bottom of a common chip box will answer very well; only surrounding the edge of it with soft wax, to keep in the water or other fluid employed in preserving the insect. The body is then to be opened; and if the parts are soft like those of a caterpillar, they should be turned back, and fixed to the trough by small pins, which ought to be set by a small pair of nippers. At the same time, the skin being stretched by another pair of sharp forceps, the insect must be put into water, and diffected therein, occasionally covering it with spirit of wine. Thus the subject will be preserved in perfection, so that its parts may be gradually unfolded, no other change being perceived than that the soft elastic parts become stiff and opaque, while some others lose their colour.

The following instruments were made use of by M. Lyonet in his diffection of the *Chenille de Saul*. A pair of scissars as small as could be made, with long and fine arms: A pair of forceps, with their ends so nicely adjusted, that they could easily lay hold of a spider's thread, or a grain of sand: Two fine steel needles fixed in wooden handles, about two inches and three quarters in length; which were the most generally useful instruments he employed.

Dr Hooke, who likewise made many microscopic obfervations, takes notice, that the common ant or pismire is much more troublesome to draw than other insects, as it is extremely difficult to get the body in a quiet natural posture. If its feet be fettered with wax or glue, while the animal remains alive, it so twirls its body, that there is no possibility of gaining a proper view of it; and if it be killed before any obfervation is made, the shape is often spoiled before it can be examined. The bodies of many minute insects, when their life is destroyed, instantly shrivel up; and this is obfervable even in plants as well as insects, the surface of these small bodies being affected by the least change of air; which is particularly the case with the ant. If this creature, however, be dropped into rectified spirit of wine, it will instantly be killed; and when it is taken out, the spirit of wine evaporates, leaving the animal dry, and in its natural posture, or at least in such a state that it may easily be placed in whatever posture we please.

Parts of Insects. The *wings*, in many insects, are so transparent, that they require no previous preparation; but some of those that are folded up under *elytra* or cases, require considerable share of dexterity to unfold them; for these wings are naturally endowed with such a spring, that they immediately fold themselves again, unless care be taken to prevent them. The wing of the earwig, when expanded, is of a tolerable size, yet is folded up under a case not one eighth part of its bulk; and the texture of this wing renders it difficult to be unfolded. This is done with the least trouble immediately after the insect is killed. Holding then the creature by the thorax, between the finger and thumb, with a blunt-pointed pin endeavour gently to open it, by spreading it over the fore-finger, and at the same time gradually sliding the thumb over it. When the wing is sufficiently expanded, separate it from the insect by a sharp

knife or a pair of scissars. The wing should be prefsed for some time between the thumb and finger before it be removed; it should then be placed between two pieces of paper, and again pressed for at least an hour; after which time, as there will be no danger of its folding up any more, it may be put between the talc, and applied to the microscope. Similar care is requisite in displaying the wings of the notonecta and other water-insects, as well as most kinds of gryllí.

The minute *scales* or *feathers*, which cover the wings of moths or butterflies, afford very beautiful objects for the microscope. Those from one part of the wing frequently differ in shape from such as are taken from other parts; and near the thorax, shoulder, and on the fringes of the wings, we generally meet with hair instead of scales. The whole may be brushed off the wing, upon a piece of paper, by means of a camel's-hair pencil; after which the hairs can be separated with the assistance of a common magnifying glass.

It is likewise a matter of considerable difficulty to diffect properly the *probofcis* of insects, such as the gnat, tabanus, &c. and the experiment must be repeated a great number of times before the structure and fituation of the parts can be thoroughly investigated, as the obferver will frequently difcover in one what he could not in another. The *collector of the bee*, which forms a very curious object, ought to be first carefully washed in spirit of turpentine; by which means it will be freed from the unctuous matter adhering to it; when dry, it is again to be washed with a camel's-hair pencil to difengage and bring forward the small hairs which form part of this microscopic beauty. The best method of managing the *flings* of insects, which are in danger of being broken by reason of their hardness, is to soak the case and the rest of the apparatus for some time in spirit of wine or turpentine; then lay them on a piece of paper, and with a blunt knife draw out the sting, holding the sheath with the nail of the finger or any blunt instrument; but great care is necessary to preserve the *feelers*, which when cleaned add much to the beauty of the object. The *head* of the lepas antifera is to be soaked in clean soft water, frequently brushing it while wet with a camel's-hair pencil: after it is dried, the brushing must be repeated with a dry pencil to difengage and separate the hairs, which are apt to adhere together.

To shew to advantage the *fat*, *brain*, and other fimilar fubfiances, Dr Hooke advifes to render the furface fmooth, by preffing it between two plates of thin glass, by which means the matter will be rendered much thinner and more transparent: without this precaution, it appears confused, by reason of the parts lying too thick upon one another. For *muscular fibres*, take a piece of the flesh, thin and dry; moisten it with warm water, and after this is evaporated the veffels will appear more plain and distinct; and by repeated macerations they appear still more so. The *crusta* of insects afford a pleafing object, and require but little preparation. If heat or curled up, they will become so relaxed by being kept a few hours in a moist atmosphere, that you may easily extend them to their natural positions; or the steam of warm water will anfwer the purpofe very well.

The *eyes* of insects in general form very curious and

3

MIC [732] MIC

II. Befides the objects for the microfcope already mentioned, there are innumerable others, fome hardly vifible, and others totally invifible, to the naked eye; and which therefore, in a more peculiar fenfe, are denominated,

Microfcopic animals. They are the animalcules or moving bodies in water, in which certain fubftances have been infufed; and of which there are a great many different kinds. Thefe animalcules are fometimes found in water which we would call pure, did not the microfcopes difcover its minute inhabitants; but not equally in all kinds of water, or even in all parts of the fame kind of it. The furfaces of infufions are generally covered with a fcum which is eafily broken, but acquires thicknefs by ftanding. In this fcum the greateft number of animalcules are ufually found. Sometimes it is neceffary to dilute the infufions; but this ought always to be done with water, not only diftilled, but viewed through a microfcope, left it fhould alfo have animalcules in it, and thus prove a fource of deception. It is, however, moft proper to obferve thofe minute objects after the water is a little evaporated; the attention being lefs diverted by a few objects than when they appear in great numbers. One or two of the animalcules may be feparated from the reft by placing a fmall drop of water on the glafs near that of the infufion; join them together by making a fmall connection between them with a pin; and as foon as you perceive that an animalcule has entered the clear drop, cut off the connection again.

Eels in pafte are obtained by boiling a little flour and water into the confiftence of book-binders pafte; then expofing it to the air in an open veffel, and beating it frequently together to keep the furface from growing mouldy or hard. In a few days it will be found peopled with myriads of little animals vifible to the naked eye, which are the eels in queftion. They may be preferved for a whole year by keeping the pafte moiftened with water; and while this is done, the motion of the animals will keep the furface from growing mouldy. Mr Baker diftils a drop or two of vinegar to be put into the pafte now and then. When they are applied to the microfcope, the pafte muft be diluted in a piece of water for them to fwim in.

Numberlefs animalcules are obferved by the microfcope in infufions of pepper. To make an infufion for this purpofe, bruife as much common black pepper as will cover the bottom of an open jar, and lay it thereon about half an inch thick; pour as much foft water into the veffel as will rife about an inch above the pepper. Shake the whole well together; after which they muft not be ftirred, but be left expofed to the air for a few days; in which times thin pellicle will be formed on the furface, in which innumerable animals are to be obferved by the microfcope.

The microfcopic animals are fo different from thofe of the larger kinds, that fcarce any fort of analogy feems to exift between them; and one would almoft be tempted to think that they lived in confequence of laws directly oppofite to thofe which preferve ourfelves and other vifible animals in exiftence. They have been fyftematically arranged by O. F. Muller; though it is by no means probable that all the different claffes

have yet been difcovered. Such as have been obferved, however, are by this author divided in the following manner.

I. *Such as have no external organs.*

1. Monas : Punctiforma. A mere point.
2. Proteus : Mutabilis. Mutable.
3. Volvox : Sphæricum. Spherical.
4. Enchelis : Cylindracea. Cylindrical.
5. Vibrio : Elongatum. Long.

* Membranaceous.

6. Cyclidium : Ovale. Oval.
7. Paramecium : Oblongum. Oblong.
8. Kolpoda : Sinuatum. Sinuous.
9. Gonium : Angulatum. With angles.
10. Burfaria. Hollow like a purfe.

II. *Thofe that have external organs.*

* Naked, or not inclofed in a fhell.

1. Cercaria : Caudatum. With a tail.
2. Trichoda : Crinitum. Hairy.
3. Kerona : Corniculatum. With horns.
4. Himantopus : Cirratum. Cirrated.
5. Leucophra : Ciliatum undique. Every part ciliated.
6. Vorticella : Ciliatum apice. The apex ciliated.

* Covered with a fhell.

7. Brachionus : Ciliatum apice. The apex ciliated.

I. *Monas.*

This is by our author defined to be " an invifible (to the naked eye), pellucid, fimple, punctiform worm;" but of which, fmall as it is, there are feveral fpecies.

1. The monas *termo* or *gelatinofa*, is a fmall jellylike point, which can be but imperfectly feen by the fingle microfcope, and not at all by the compound one. In a full light they totally difappear, by reafon of their tranfparency. Some infufions are fo full of them that fcarce the leaft empty fpace can be perceived; the water itfelf appearing compofed of innumerable globular points, in which a motion may be perceived fomewhat fimilar to that which is obferved when the fun's rays fhine on the water; the whole multitude of animals appearing in commotion like a hive of bees. This animal is very common in ditch-water, and in almoft all infufions either of animal or vegetable fubftances.

2. Monas *atomus* or *atilda*; white monas with a variable point. This appears like a white point, which thro' a high magnifier appears fomewhat egg-fhaped. The fmaller end is generally marked with a black point, the fituation of which is variable; fometimes it appears on the large end, and fometimes there are two black fpots in the middle. This fpecies was found in fea-water, which had been kept through the whole winter, but was not very fetid. No other kind of animalcule was found in it.

3. Monas *punctum* or *nigra*, black monas. This was found in a fetid infufion of pears, and appears in form of a very minute, opaque, and black point, moving with a flow and wavering motion.

4. Monas *ocellus*, tranfparent like talc, with a point in the middle. This is found in ditches covered with confervæ.

4

Microscopes, and sometimes with the cyclidium milium; the margin of it is black, with a black point in the middle.

5. *M. star her or pratiser of a tiery appearance.* This is found in all kinds of water; sometimes even in that which is pure, but always in the commun-time in ditch-water. It is found also in infusions of animal or vegetable substances, whether in fresh or salt water; myriads being contained in a single drop. It is found likewise in the film of the teeth. It is nearly of a round figure; and so transparent, that it is impossible to discover the least vestige of intestines. They generally appear in clusters, but sometimes singly. Contrary to what happens to other animalcules, they appear to cover the edges of the drop when evaporating, and where they constantly die. A few dark shades, probably occasioned by the wrinkling of the body, are perceived when the water is nearly evaporated. The motions of this animalcule are generally very quick; and two united together, may sometimes be seen swimming among the reft; which is thought to be a single one generating another by division, as is related under the article ANIMALCULE. These and the animalcules of the first species are so numerous, that they exceed all calculation even in a very small space.

6. *M. marked* with a circle. This is found in the purest waters, and may be discovered with the third lens of the single microscope when the magnifying power is increased. It appears like a small lucid point; but can assume an oval or spherical shape at pleasure: sometimes the appearance of two kidneys may be perceived in its body, and there is commonly the figure of an ellipse in it; the situation of which is moveable, sometimes appearing in the middle and sometimes approaching to either extremity. It seems encompassed with a beautiful halo, which is thought to be occasioned by the vibration of fine invisible hairs. It has a variety of motions, and often turns round for a long time in the same place.

7. *The tranquilla,* or egg-shaped transparent mons with a black margin, is found in urine which has been kept for some time. Urine in this state acquires a form in which the animalcules reside; but though kept for several months, no other species was found in it. A drop of urine is unsafe fatal to other animalcules, though this species is to be met with in no other substance. It is generally fixed to one point, but has a kind of oscillatory motion. Frequently these creatures are surrounded with a halo. Sometimes they are quadrangular, and at other times spherical; the black margin is not always to be found; and sometimes there is even no appearance of a tail.

8. *The lenticula,* or flat transparent mosss, is most usually found in salt-water; is of a whitish colour and transparent, more than twice as long as it is broad, with a dark margin, having a vacillatory motion, and frequently appearing as double.

9. *The polyzoaria* or mons with a green margin. These are generally found in marshy grounds in the month of March. They appear like small spherical grains of a green colour on the circumference, having sometimes a green bent line passing through the middle. They appear sometimes in clusters, from three to seven or more in number, having a wavering kind of motion.

10. *The mons,* or transparent gregarious mons, is found in a variety of infusions, and is of that kind which multiplies by dividing itself. They appear in clusters of four, five, or sometimes many more; the corpuscles being of various sizes, according to the number collected into one group. The smaller particles, when separated from the larger, move about with incredible swiftness. A single corpuscle separated from the heap, and put by itself into a glass, soon increased in size till it nearly attained the bulk of the parent group. The surface then assumed a wrinkled appearance, and gradually became like the former, separating again into small particles, which likewise increased in bulk as before.

II. *The Proteus.*

An invisible, very simple, pellucid worm, of a variable form.

1. The *diffluens,* branching itself out in a variety of directions. It is very rare, and only met with in fens; appearing like a grey mucous mass, filled with a number of black globules, and continually changing its figure, pushing out branches of different lengths and breadths. The internal globules divide immediately, and pass into the new formed parts; always following the various changes of the animalcule; which changes seem to proceed entirely from the internal mechanism of its body, without the aid of any external power.

2. The *tenax,* running out into a fine point. This is a pellucid gelatinous body, stored with black molecules, and likewise changing its figure, but in a more regular order than the former. It first extends itself in a straight line, the lower part terminating in a bright acute point. It appears to have no intestines; and when the globules are all collected in the upper part, it next draws the pointed end up toward the middle of the body, which assumes a round form. It goes through a number of different shapes, part of which are described under the article ANIMALCULE. It is found in some kinds of river-water, and appears confined almost entirely to one place, only bending likewise.

III. *Volvox.*

An invisible, very simple, pellucid, spherical worm.

1. The *punctum p* of a black colour, with a lucid point. This is a small globule, with one hemisphere opaque and black, the other having a crystalline appearance; and a vehement motion is observed in the black part. It moves as on an axis, frequently passing thro' the drop in this manner. Many are often seen joined together in their passage through the water; sometimes moving as in a little whirlpool, and then separating. They are found in great numbers on the surface of fetid sea-water.

2. The *granulum* is of a spherical figure and green colour, the circumference being bright and transparent. It is found in marshy places about the month of June, and moves but slowly. It seems to have a green opaque nucleus.

3. The *globulus,* with the hinder part somewhat obscure, sometimes verges a little towards the oval in its shape, having a slow fluttering kind of motion, but

more

Microscope more quick when disturbed. The intestines are but just visible. It is found in most vegetable infusions, and is ten times larger than the æmma lens.

4. The *pilula*, small and round, with green intestines. This is found in water where the lemna minor grows, in the month of December, and has a kind of rotatory motion, sometimes slow and at others quick. The intestines are placed near the middle, apparently edged with yellow. There is a small incision on one of the edges of the sphere, which may possibly be the mouth of the creature. The whole animal appears encompassed with an halo.

5. The *grandinella*, with immoveable intestines, is much smaller than the last, and marked with several circular lines. The intestines are immoveable, and no motion is perceived among the interior molecules. Sometimes it moves about in a straight line, at others irregularly, and sometimes keeps in the same spot, with a tremulous motion.

6. The *sociale*, with crystalline molecules placed at equal distances from one another. This is found in water where the chara vulgaris has been kept; and has its molecules disposed in a sphere, filling up the whole body of the animalcule; but whether they be covered by a common membrane or united by a stalk (as in the *vorticella sociale* to be afterwards described) is not known. When very much magnified, some black points may be seen in the crystalline molecules. Its motion is sometimes rotatory and sometimes not.

7. The *sphærieula*, with round molecules, appears to consist of pellucid homogeneous points of different sizes. It moves slowly from right to left and back again, about a quarter of a circle each time.

8. The *lunula*, with lunular molecules, is a small roundish transparent body, consisting of an innumerable multitude of homogeneous molecules of the shape of a crescent, without any common margin. It moves occasionally in a twofold manner, viz. of the molecules among one another, and the whole mass turning slowly round. It is found in marshy places in the beginning of spring.

9. The *globator*, or spherical membranaceous valves, is found in great numbers in the infusions of hemp and nettles, and in stagnant waters in spring and summer; it was first observed and depicted by Leeuwenhoeck, but the descriptions of it given by authors differ considerably from each other. The following is that of Mr Baker. "There is no appearance of either head, tail, or feet. It moves in every direction, backwards, forwards, up or down, rolling over and over like a bowl, spinning horizontally like a top, or gliding along smoothly without turning itself at all; sometimes its motions are very slow, at other times very swift; and when it pleases it can turn round as upon an axis very nimbly, without moving out of its place. The body is transparent, except where the circular spots are placed, which are probably its young. The surface of the body in some is as if all dotted over with little points, and in others as if granulated like fine green. In general it appears as if set round with those moveable hairs." Another author informs us, that "they are at first very small, but grow so large that they can be discerned with the naked eye; they are of a yellowish green colour, globular figure, and in

substance membranaceous and transparent; and in the middle of this substance several small globes may be perceived. Each of these are smaller animalcules, which have also the diaphanous membrane, and contain within themselves still smaller generations, which may be distinguished by means of very powerful glasses. The larger globules may be seen to escape from the parent, and then increase in size."

This little animal appears like a transparent globule of a greenish colour, the fœtus being composed of smaller greenish globules. In proportion to its age it becomes whiter and brighter, and moves slowly round its axis; but to the microscope its surface appears as if granulated, the roundest molecules fixed in the centre being largest in those that are young. The exterior molecules may be wiped off, leaving the membrane naked. When the young ones are of a proper size, the membrane opens, and they pass through the fissure; after which the mother turns away. Sometimes they change their spherical figure, and become flat in several places. They contain from 8 to between 30 and 40 globules within the membrane.

10. The *æreon*, with spherical green globules in the centre. This is found amongst the lemna in the months of October and December, and has a slow rotatory motion. The globules seldom move, though a slow quivering motion may sometimes be perceived among them in the centre.

11. The *alten*, composed of green globules not inclosed in any membrane, is found in the month of August in water where the lemna polyrrhiza grows. —It consists of a congeries of greenish-coloured globules, apparently of an equal size, with a bright spot in the middle; the whole mass is sometimes of a spherical form, sometimes oval, without any common membrane: a kind of halo may be perceived round it, and the mass generally moves from right to left, but scarce any motion can be perceived among the globules themselves. These masses contain from four to fifty globules, of which a solitary one may sometimes be seen. Sometimes also two masses of globules have been perceived joined together.

12. The *vegetans*, terminating in a little bunch of globules. This is found in river-water in the month of November. It consists of a number of floccula opaque branches invisible to the naked eye; and at the apex of these is a small congeries of very minute oval pellucid corpuscles. Müller, who discovered this, supposed it at first to be a species of microscopic and river sertularia; but he afterwards found the branches quitting the branches, and swimming about in the water with a proper spontaneous motion; many of the old branches being deserted, and the younger ones furnished with them.

IV. *Enchelis*:
A simple, invisible, cylindric worm.

1. The *viridis*, or green enchelis, has an obtuse tail, the forepart terminating in an acute truncated angle; the intestines are obscure and indistinct. It continually varies its motion, turning from right to left.

2. The *punctifera*, having the fore part obtuse, the hinder part pointed. It is opaque, and of a green colour, with a small pellucid spot in the fore part, in which two black points may be seen; and a kind of double

microscope double band crosses the middle of the body. The hinder part is pellucid and pointed, with an incision, supposed to be the mouth, at the apex of the fore-part. It is found in marshes.

3. The *dotes*, or gelatinous enchelis, is found, though rarely, in an infusion of lemon, and moves very slowly. The body is round, of a very dark-green, the fore-part bluntly rounded off, and the hinder part somewhat tapering, but finished with a round end: near the extremities there is a degree of transparency.

4. The *finalis*, with moveable intestines, is found in water that has been kept for several months: it is of an egg-shape, and generally moves very quick, either to the right or left. It is supposed to be furnished with hairs, because when moving quickly the margin appears fringed. The body is opaque with a pellucid margin, and filled with moveable spherules.

5. The *firetina*, with immoveable intestines, is of an oval figure, partly cylindrical, the fore-part smaller than the hind, with a black margin, full of gray vesicular molecules: it moves very slowly.

6. The *edulosa*, with visible moveable intestines, is found in the same water with the cyclidium glaucoma, but is much more scarce. The body is egg-shaped, the fore-part narrow, and frequently filled with opaque confused intestines: when moving, it elevates the fore-part of the body. It is about three times as large as the cyclidium glaucoma.

7. The *fontinalum* is found in water that has been kept for some days, and moves by ascending and descending alternately. It is of a cylindrical figure, twice as long as broad, the intestines in the fore-part transparent, but opaque in the hinder part. Sometimes it is observed swimming about with the extremities joined together.

8. The *intermedia*, with a blackish margin, is one of the smallest animalcules: it has a transparent body, without any visible intestines. The fore and hind parts are of an equal size, and the edge is of a deeper colour than the rest. Some have a point in the middle, others a line passing through it.

9. The *firetina*, is transparent, round, and egg-shaped. A very strong magnifier discovers some long foldings on the surface, with a few bright molecules here and there.

10. The *pirum*, with the hinder part transparent, has the fore-part protuberant and filled with molecules. The hinder part is smaller and empty, with moveable molecular intestines. Its motion is rapid, passing backwards and forwards through the diameter of the drop. When at rest, it appears to have a little swelling on the middle of the body.

11. The *tremula* was found in an infusion with the paramoecia aurelia, and many other animalcules. It is among the least of these minute creatures, and is of a cylindrical figure and gelatinous texture. Its extremity appears pointed, and has a tremulous motion, so as to induce a suspicion that the creature has a tail. Two of these creatures may at times be seen to adhere together.

12. The *confinis*, with a stricture in the middle, is found in lake-water, and is of a very small size, having the middle drawn in as if tied with a string. It is of an oval shape.

13. The *olipins*, with a congeries of green intestines, is found among the green matter on the sides of vessels in which water has been kept for some time. It is of a roundish shape, and transparent; the fore-part obtuse, the hinder part rather sharp, and marked with green spots. They are generated in such numbers, that myriads may sometimes be found in one drop.

14. The *febris*, with both ends truncated, was found in water called pure, and had a languid motion. The body is round and transparent, with the fore and hind parts somewhat smaller than the rest. In the inside is a long and somewhat winding intestine, with a bright sky-coloured fluid, and some black molecule transversely situated.

15. The *fruticula*, with the fore-part truncated, is found in an infusion of grass and hay, and runs backward and forward through the drop with a wavering motion. It is one of the most transparent animalcules, and has the fore-part obtusely convex.

16. The *caudata*, with a kind of tail, is but seldom met with. The body is grey and transparent, with globular molecules divided from each other, and dispersed thro' the whole; the fore-part is thick and obtuse, the hind part crystalline and small, the end truncated.

17. The *epifonium*, with the fore-part slender and roundish, is among the smaller animalcula; the body cylindrical and bright, the hind part obtuse, the fore-part smaller, and terminating in a globule, with now and then a black line down the middle.

18. The *gemmata* is found in ditch-water where the lemna thrives. The upper part running out into a transparent neck, with a double series of globules running down the body. It moves slowly, and generally in a straight line.

19. The *retrograda* moves commonly sideways, and sometimes in a retrograde manner. It has a gelatinous transparent body, thicker in the middle than at the ends, without any thing that can be called intestines, except a pellucid globule discoverable near the hinder part.

20. The *fylifaver*, with obtuse ends, is found in sea-water, and has a quick oscillatory motion from one side to the other. The body is round, with the fore-part transparent. More than half the length of it is without any visible intestines; but the lower end is filled with minute vesicular and transparent globules; a large globular vesicle is also observed in the fore-part.

21. The *furcinum* was found by Joblot in an infusion of blue bottles, moving very slowly in an undulatory manner. The body is cylindrical, about four times as long as broad, truncated at both ends, the intestines opaque, and not to be distinguished from one another. It turns itself into the shape of the letter S, by turning the two extremities contrariwise.

22. The *index* is found in water with the lemna minor: the body opaque, of a grey colour, and long conical shape; the lower end is obtuse, one side projecting like a finger from the edge, with two very small projections from the lower end. It has the power of retracting these projections, and making both ends appear obtuse.

23. The *tremex*, with a kind of beak, is the largest of this kind of animalcules. The body is grey, long, and mucous; the fore-part globular, the hinder part obtuse; but it can alter its shape considerably. Sometimes there is an appearance of three teeth proceeding from

Microscope from one of the fides. Globules of different sizes may be observed within the body. The creature rolls slowly about from right to left.

14. The *larva* is long, round, and filled with molecules. The fore-part is obtufe and tranfparent, with a kind of neck or fmall contraction formed near this end: the lower part is pointed; and about the middle of the body are two fmall pointed projections like nipples, one on each fide.

15. The *fpatula*, with the fore-part tranfparent, and of the fhape of a fpatula. It is perfectly cylindrical, cryftalline, and marked with fine longitudinal furrows; having generally two tranfparent globules, one below the middle, the other near the extremity. It moves in a wavering kind of manner, retaining its general form, but moving the fpatula in various ways. Muller informs us, that he faw it once draw the fpatula within the body, and keep it there for two hours.

16. The *papula*, with the fore-part papillary, is found in dunghill water in November and December: it has a rotatory motion on a longitudinal axis, and moves in an oblique direction through the water. Both ends are obtufe; and the hinder part is marked with a tranfparent circle, or circular aperture.

17. The *pupa*, with a fmall nipple proceeding from the apex, has a very flow motion, and refembles the former, only that it wants the tranfparent circle, and is much larger. It is all opaque but the forepart, and filled with obfcure points.

V. *Vibrio:*
A very fimple, invifible, round, and rather long worm.

1. The *lineola* is found in moft vegetable infufions in fuch numbers, that it feems to fill up almoft the whole of their fubftance. It is fo fmall, that with the beft magnifiers we can difcern little more than as obfcure tremulous motion among them. It is more flender than the monas terms.

2. The *rugula* is like a bent line; and fometimes draws itfelf up in an undulated fhape, at others moves without bending the body at all.

3. The *bacillus*, equally truncated at both ends, is found in an infufion of hay; but Muller mentions the following remarkable fact, viz. that having made two infufions of hay in the fame water, he put the hay whole in the one, but cut it in pieces in the other: he found in the former none of the vibrio bacillus, but many of the monas lens and kolpoda cuculus; in the latter were many of the vibrio, but few of the other.—This is from fix to ten times longer than the monas lens, but much more flender.

4. The *undula*, is a round, gelatinous, little, undulating line. This is the animal which Leewenhoeck fays is lefs than the tail of one of the feminal animalcules. It never appears ftraight; but when at reft it refembles the letter V, and when in motion the letter M. It commonly refts on the top of the water; fometimes it fixes itfelf by one extremity, and whirls round.

5. The *ferpens*, with obtufe windings or flexures, is found in river-water, but feldom. It is flender, and gelatinous, refembling a ferpentine line, with an inteftine down the middle.

6. The *fpirillum* is exceedingly minute, and twifted in the form of a fpiral, which feems to be its natural

Nº 219.

fhape, as it never unttwifts itfelf, but moves forward in a ftraight line, vibrating the hind and fore parts. It was found in 1782 in an infufion of the fonchus arvenfis.

7. The *vermiculus* has a milky appearance, with an obtufe apex, and a languid undulatory motion, like that of the common worm. It is found in marfhy water in November, but feldom. It is thought to be the animal mentioned by Leewenhoeck as found in the dung of the frog and fpawn of the male libellula.

8. The *inteftinum* is found in marfhy waters, and has a flow progreffive motion. It is milk-coloured, with two obtufe ends, and four or five fpherical eggs are perceivable at the hinder extremity.

9. The *bipunctatus* is found in fetid falt-water, and moves flowly; for the moft part in a ftraight line. The body is pellucid, and of a talc-like appearance; both ends are truncated, and in the middle one or two pellucid globules placed lengthwife.

10. The *tripunctatus* is alfo tranfparent and talcy, with both ends tapering. It has three pellucid globules, the middle one of which is largeft, the fpace between them being generally filled with a green matter. It moves in a ftraight line, backwards and forwards.

11. The *paxillifer*, or ftraw-like vibrio, confifts of a tranfparent membrane, with yellow inteftines, and two or three vifible points. They are found in parcels together from feven to forty in number, and ranged in a variety of forms. When at reft, they generally affume a quadrangular figure; and are thought to have fome affinity to the bair-like animal defcribed by Mr Baker, and of which an account is given under the article ANIMALCULE, nº 3.

12. The *lunula*, or bow-fhaped vibrio, refembles the moon at its firft quarter; it is of a green colour, and has from feven to ten globules difpofed in a longitudinal direction.

13. The *vermiluus* is found in great plenty in falt-water kept for fome days till it becomes fetid. It moves quickly, and with an undulatory motion, backwards and forwards. It is a long tranfparent membrane, with the hind part broader than the fore one. Thefe animalcula feem to be joined together in a very fingular manner.

14. The *mallous* is found in great plenty in fpring-water, and is alternately at reft and in motion every moment; in the former cafe refembling the letter T, and in the latter V. It is a white pellucid animalcule, with a globule affixed to the bafe.

15. The *acus* is in the fhape of a fewing needle; the neck round and partly tranfparent, and marked in the middle with a red point; the tail refembling a fine briftle.

16. The *fagitta*, with a fetaceous tail, has a long and flexible body; broadeft about the middle, and filled there alfo with grey molecules; the fore-part being drawn out into a thin and tranfparent neck, and the upper end thick and black. It is found in falt-water, and feems to move by contracting and extending its neck.

17. The *gordius*, with a tail terminated by a fmall tubercle, was found in an infufion made with falt water. Its fore part throughout about one fixth of its length is tranfparent, and furnifhed with an alimentary tube of a fky colour; the lower part being

being bright and pointed, and the middle full of small globules.

18. The *serpentulus*, somewhat pointed at both ends. This is found in the infusions of vegetables which have been kept for some weeks. Its body is of a whitish colour, frequently convoluted, and drawn into different figures. The tail is furnished with a long row of very minute points.

19. The *coluber* is found in river water; the tail is extremely small, and bent so as to form a considerable angle with the body; the mouth, œsophagus, the molecules in the intestines, and the coilings of them, are easily discerned.

20. The *anguillula* is divided into four varieties: 1. The vinegar eel; 2. That in paste; 3. That of fresh water; and, 4. That of salt. The two first are treated of under the article ANIMALCULE; the third is exceedingly transparent, with a few transverse lines upon the body, but without any appearance of intestines. Sometimes it has a long row of little globules, and is frequently furnished with two small oval ones; the tail terminates in a point. It has been found in the sediment formed by vegetables on the sides of vessels in which water had been kept for a long time. The fourth variety appears, when pressed between two glass plates, to be little more than two crystalline skins with a kind of intestines of a clay colour. The younger ones are furnished with pellucid molecules.

21. The *luctor*, or ventricose oval vibrio, with a short neck, is found among the larger kinds of animalcules, egg-shaped, pellucid, inflated, and somewhat depressed at top; having a moveable crystalline neck, and the belly filled with pellucid molecules.

22. The *utriculus* resembles a bottle; the belly is full of molecular intestines, the neck bright and clear, the top truncated, and some have a pellucid point at the bottom of the belly. It has a constant and violent vacillatory motion, the neck moving very quickly from side to side.

23. The *fasciola* is found in water just freed from the frost, and not often in any other fluid. It is pellucid, with intestines like points in the middle. There is likewise an alimentary canal gradually diminishing in size. Its motion is very quick.

24. The *colymbus* is larger than many of the other species of vibrio, and resembles a bird in shape. The neck, which is a little bent, is round, shorter than the trunk, of an equal size throughout, and of a bright appearance, with the apex obtuse. The trunk is thick, somewhat triangular, full of yellow molecules; the fore-part broad, the hinder part acute, the motion slow.

25 The *fluvius* has a linear body, being a bright membranaceous thread; the hinder part somewhat thicker, round, and filled with molecules, excepting at the end, where there is a small empty pellucid space. It can draw in the slender filiform part at pleasure.

26. The *cauda*, with both ends attenuated, and the neck longer than the tail, is found in salt water; tho' a kind is likewise found in fresh water with a neck longer than the other. The trunk of this animalcule is oblong, opaque, and filled with molecules; the fore and hind parts are drawn out into a pellucid taley membrane, which the creature can retract at pleasure.

VOL. XI. Part II.

27. The *cygnus* is a very pellucid line, crooked at top, swelling in the middle, and sharp at the end; the middle full of dark coloured molecules and pellucid intestines. It is very small, and moves more slowly than any of those that move and advance their necks.

28. The *anser* is found in water where duckweed grows. The trunk is elliptic, round, and without any inequality on the sides. It is full of molecules; the hind part sharp and bright; the fore part produced into a bending neck, longer than the body; the apex whole and even, with blue canals passing between the marginal edges, occupying the whole length of the neck; and in one of them a violent descent of water to the beginning of the trunk is observable. It moves the body slow, but the neck more briskly.

29. The *olor* is found in water that has been kept for a long time, and is full of vegetable green matter. The body is elliptical and ventricose, the hind part somewhat sharp, and sometimes filled with darkish molecules. The neck is three or four times longer than the body; of an equal size throughout, and is moved very quickly; but the motion of the body itself is slow.

30. The *fala*, with a crooked neck, and obtuse hinder part, is pellucid and elliptical; the fore part fastening into a little, round, bright neck, nearly as long as the trunk. The latter is somewhat gibbous, and filled with very small molecules; and there are two small bright globules, one within the hind extremity, and the other in the middle of the body. The neck of this animalcule is immoveable; whence it moves something like a scythe.

31. The *intermedius* appears to be an intermediate species betwixt the fala and the fasciola. It seems to be a thin membrane constantly folded. The whole has a crystalline talcy appearance; the middle filled with grey particles of different sizes. It has all round a distinct bright margin.

VI. *Cyclidium*.

A simple, invisible, flat, pellucid, orbicular or oval worm.

1. The *bulla*, or orbicular bright cyclidium. This is found occasionally in an infusion of hay. It is very pellucid and white, but the edges somewhat darker than the rest. It moves slowly, and in a semicircular direction.

2. The *milium* is very pellucid and splendid like crystal; and of an elliptical figure, with a line through the whole length of it. The motion is swift, interrupted, and fluttering.

3. The *flexuum* is one of the smallest animalcula; the body somewhat of an oval shape, with two small blue spaces at the sides.

4. The *glutinum* has an oval pellucid body, with both ends plain, or an oval membrane with a distinct well-defined edge. The intestines are so transparent, that they can scarce be discerned when it is empty. When full, they are of a green colour, and there are dark globules discoverable in the middle. When there is plenty of water this animalcule moves swiftly in a circular and diagonal direction; when it moves slowly, it seems to be taking in water, and the intestines are in a violent commotion. It generates by division.

5. The *equinum* is very small, pellucid, and flat, with a black margin.

5 A   6. The

Microscope

6. The *restratum* is oval, smooth, and very pellucid, with the fore part running out into an obtuse point, with which it seems to feel and examine the bodies to which it comes. The intestines are filled with a blue liquor, the colour of which sometimes vanishes, and then they seem to be composed of vesicles.

7. The *nucleus* resembles a grape seed, the body being pellucid and depressed, the fore part obtusely convex, and the hind part acute.

8. The *lyatinum* has a tremulous kind of motion; the body oval, flat, and bright, without any visible intestines.

9. The *pediculus* is scarce ever seen but on the hydan pallida, upon which it runs as if it had feet. It is gelatinous and white; the bottom gibbous over the back; the extremities depressed and truncated, with one end sometimes apparently cloven into two, which may be supposed the mouth.

10. The *dolium* is of an oval shape, with one side convex, the other concave; the margin pellucid, and the inner part containing a great number of molecules.

VII. *Paramecium.*
An invisible, membranaceous, flat, and pellucid worm.

1. The *aurelia* is membranaceous, pellucid, and four times longer than it is broad; the fore part obtuse and transparent; the hind part filled with molecules. It has somewhat the appearance of a gimlet by reason of a fold which goes from the middle to the apex, and is of a triangular figure. It moves in a rectilineate and vacillatory manner. It is found in ditches where there is plenty of duckweed, and will live many months in the same water without any renewal of the latter.

2. The *chrysalis* is found in salt water, and differs very little from the former, only the ends are more obtuse, and the margins are filled with black globules.

3. The *versutum* is found in ditches, and has an oblong, green, and gelatinous body, filled with molecules; the lower part thicker than the other; and both ends obtuse. It propagates by division.

4. The *onsersum* is membranaceous, oval, grey, and pellucid, with many oval corpuscles dispersed through the body.

5. The *marginatum* is flat, elliptical, and every where filled with molecules, except in the lower end where there is a pellucid vehicle. It is surrounded by a broad double margin, and a bright spiral intestine is observable.

VIII. *Kolpoda.*
An invisible, pellucid, flat, and crooked worm.

1. The *lamella* is very seldom met with. It resembles a long, narrow, and pellucid membrane, with the hind part obtuse, narrower, and curved towards the top. It has a vacillatory and very singular motion; going upon the sharp edge, as on the flat side as is usual with microscopic animals.

2. The *gallinula* is found in fetid salt water; and has the apex somewhat bent, the belly oval, convex, and pointed.

3. The *rostrum* is found though seldom, in water where the lemna grows; and has a slow and horizontal motion. The fore part is bent into a kind of hook; the hind part obtuse, and quite filled with black molecules.

4. The *ochrea* is depressed, membranaceous, and flexible; one edge nearly straight; the other somewhat bent, filled with obscure molecules, and a few little bladders dispersed here and there.

5. The *mucronata* is a dilated bright membrane; the apex an obtuse point, with a broad marked border running quite round it. It is filled with grey molecules within the margin, and has a truncated appearance.

6. The *triquetra* was found in salt water, and appears to consist of two membranes; the upper side flattened, the lower convex, with the apex bent into a kind of shoulder.

7. The *fixata* is likewise found in salt water, and is very pellucid and white, with the upper part rather bent, and terminating in a point; the lower part obtusely round; there is a little black pellucid vehicle at the apex; and with a very great magnifying power the body appears covered with long streaks.

8. The *cuculus* is of an oval shape, with the vertex pointed, and of a brilliant transparency, by which the viscera are rendered visible. These consist of a number of round diaphanous vesicles.

9. The *meleagris* has a dilated membrane, with very fine folds, which it varies in a moment. The fore part of the body to the middle is clear and bright; the hind part variously folded in transverse and elevated plaits and full of molecules. Beneath the apex are three or four teeth; but in some the edge is obtusely notched, and set with smaller notches. In the hinder part are 12 or more equal pellucid globules.

10. The *assimilis* is found on the sea-coast, and has an elliptic mass in the middle, but is not folded like the former. The margin of the fore part is notched from the top to the middle; the lower part swells out, and contracts again into a small point.

11. The *cucullus* is found in vegetable infusions, and in fetid hay; moving in all directions, and commonly with great vivacity. It is very pellucid, and has a well defined margin, filled with little bright vesicles differing in size, and of no certain number. Its figure is commonly oval, with the top bent into a kind of beak, sometimes oblong, but most commonly obtuse. It has in the inside from 8 to 24 bright little vesicles not discernible in such as are young. Some have supposed these to be animalcules which this creature has swallowed; but Mr Müller is of opinion that they are its offspring. When this creature is near death by reason of the evaporation of the water, it protrudes its offspring with violence. From some circumstances it would seem probable that this animalcule casts its skin, as is the case with some insects.

12. The *cucullulus* is found in an infusion of the sonchus arvensis. It is very pellucid and crystalline, with several globules, and has an oblique incision a little below the apex.

13. The *cuneus* is elliptical, flat on the upper side, and convex on the under; the fore part is clear, and from the middle to the hinder part is full of silver-like globules. It frequently stretches out the fore part, and folds it in different positions.

14. The

14. The *res*, or *crassa*, is found in an infusion of hay, commonly about 13 hours after the infusion is made, and has a quick and vacillatory motion. Its body is yellow, thick, and somewhat opaque; curved a little in the middle, so that it resembles a kidney; and full of molecules. When the water in which it swims is about to fail, it takes an oval form, is compressed, and at last bursts.

15. The *pirum* has an uniform and transparent body, without any sensible inequality: and is of a pale colour, with obscure little globules. It propagates by division.

16. The *cuneus* is white, gelatinous, and without any distinct viscera; having a bright striated pellucid pustule on one side of the fore part. The apex has three or four teeth; and it can bend the hinder part into a spiral form.

IX. *Gonium.*
An invisible, simple, smooth, and angular worm.

1. The *pectorale* is found in pure water, and moves alternately towards the right and left. It is quadrangular and pellucid, with 16 spherical molecules of a greenish colour, " set in a quadrangular membrane, like the jewels in the breast-plate of the high-priest, reflecting light on both sides."

2. The *pulvinatum* is found in dunghills; and appears like a little quadrangular membrane, plain on both sides: but with a large magnifier it appears like a bolster formed of three or four cylindric pillows sunk here and there.

3. The *corrugatum* is found in various kinds of infusions; and is somewhat of a square shape, very small, and in some positions appears as streaked.

4. The *rectangulum* differs but little from the former: the angle at the base is a right one; the larger vesicle is transparent, the rest green.

5. The *truncatum* is found chiefly in pure water, and then but seldom. It has a languid motion, and is much larger than the foregoing. The fore part is a straight line, with which the sides form obtuse angles, the ends of the sides being united by a curved line. The internal molecules are of a dark green, and there are two little bright vesicles in the middle.

X. *Bursaria.*
A very simple, hollow, membranaceous worm.

1. The *truncatella* is visible to the naked eye; white, oval, and truncated at the top, where there is a large aperture descending towards the base. Most of them have four or five yellow eggs at the bottom. They move from left to right, and from right to left; ascending to the surface in a straight line, and sometimes rolling about while they descend.

2. The *hirundo* is pellucid and crystalline, having splendid globules of different sizes swimming about with it. The under side is convex, the upper hollow, with the fore part forming a kind of lip.

3. The *hirundinella* has two small projecting wings, which give it somewhat of the appearance of a bird; and it moves something like a swallow. It is invisible to the naked eye; but by the microscope appears a pellucid hollow membrane.

4. The *duplella* was found among duckweed, and appears like a crystalline membrane folded up, without any visible intestines except a small congeries of points under one of the folds.

5. The *globina* has a roundish shape, and is hollow; the lower end being furnished with black molecules of different sizes, the fore part with obscure points, the rest entirely empty, and the middle quite transparent. It moves very slowly from right to left.

XI. *Cercaria.*
An invisible transparent worm with a tail.

1. The *gyrinus* greatly resembles the spermatic animalcules. It has a white gelatinous body; the fore part somewhat globular; the hind part round, long, and pointed. Sometimes it appears a little compressed on each side. When swimming it keeps its tail in continual vibration like a tadpole.

2. The *gibba* is found in the infusions of hay and other vegetables; and is small, opaque, gelatinous, white, and without any visible intestines.

3. The *inquieta* is found in salt water, and is remarkable for changing the shape of its body: sometimes it appears spherical, sometimes like a long cylinder, and sometimes oval. It is white and gelatinous, the tail thin, form and flexible, the upper part vibrating violently. A pellucid globule may be observed at the base, and two very small black points near the top.

4. The *lunus* varies its form so much, that it might be mistaken for the proteus of Baker, described under the article ANIMALCULE; though in fact it is totally different. The body sometimes appears of an oblong, sometimes of a triangular, and sometimes of a kidney shape. The tail is generally short, thick, and articulated; but sometimes long, flexible, cylindric, and without rings; vibrating, when stretched out, with so much velocity, that it appears double. A small pellucid globule, which Müller supposes to be its mouth, is observable at the apex; and two black points not easily discovered, he thinks, are its eyes. Sometimes it draws the tail entirely into the body. It walks slowly after taking three or four steps, and extends the tail, erecting it perpendicularly, shaking and bending it; in which state it very much resembles a leaf of the lemna.

5. The *turbo*, with a tail like a bristle, is found among duckweed. It is of a talcy appearance, partly oval and partly spherical; and seems to be composed of two globular bodies, the lowermost of which is the smallest, and it has two little black points like eyes on the upper part. The tail is sometimes straight, sometimes turned back on the body.

6. The *pediasus* is found in November and December, in marshy places covered with lemna. It is pellucid; and seems to consist of a head, trunk, and tail: the head resembles that of a herring; the trunk is ventricose and full of intestines, of a spiral form and black colour. The tail most commonly appears to be divided into two bristles. The intestines are in a continual motion when the body moves, and by reason of their various shades make it appear very rough. There are likewise some hairs to be perceived. It turns round as upon an axis when it moves.

7. The *viridis* is found in the spring in ditches of standing water; and in some of its states has a considerable resemblance to the last, but has a much greater power of changing its shape. It is naturally cylindrical,

Microscope lindrical, the lower end sharp, and divided into two parts; but sometimes contracts the head and tail so as to assume a spherical figure.

8. The *saifera* is found in salt water, but seldom. It is small, the body rather opaque, and of a round figure. The upper part is bright, and smaller than the rest: the trunk is more opaque; the tail sharp, and near it a little row of short hairs. It has a slow rotatory motion.

9. The *hirta* was likewise found in salt water. It is opaque and cylindrical; and when in motion, the body appears to be surrounded with rows of small hairs separated from each other.

10. The *cranona* has a ventricose, cylindrical, thick, and wrinkled body; the lower part small; the upper part terminating in a small strait neck, like that of a pitcher; the tail linear, and terminating in two diverging points.

11. The *anifer* has a moveable head fixed to the body by a point. The abdomen is twice as long as the head, full of intestines, and has a tail still narrower, and terminating in two brittles which it can unite and separate at pleasure. It moves briskly, but without going far from its first place.

12. The *cerebus* was found in a ditch where there was plenty of duckweed. It is larger than the preceding, and has a thicker and more cylindrical body; the lower part truncated, with two short diverging points projecting from the middle.

13. The *lupus* is found in water among duckweed, and is larger than most of the genus. The head is larger than the body; the apex turned down into a little hook; the tail is like the body, but narrower, terminating in two very bright spines, which it extends in different directions. Sometimes it contracts into one half its common size, and again extends itself as before.

14. The *vermicularis* is long, cylindrical, fleshy, and capable of changing its shape. It is divided into eight or nine rings or folding plaits; the apex either obtuse, or notched into two points; the hinder part rather acute, and terminating in two pellucid thorns, between which a feeling is sometimes perceived. It often projects a kind of cloven proboscis from the incision at the apex. It is found in water where there is duckweed.

15. The *forcipata* is found in marshy places, is cylindrical and wrinkled, with a forked proboscis which it can thrust out or pull in.

16. The *pleuronectes* is found in water which has been kept for several months. It is membranaceous, roundish, and white, with two blackish points in the fore part, the hinder part being furnished with a slender sharp tail. It has orbicular incisures of different sizes in the middle; the larger of them height. The motion is vacillatory; and in swimming it keeps one edge of the lateral membrane upwards, the other folded down.

17. The *tripos* is flat, pellucid, triangular, having each angle of the base or fore part bent down into two linear arms, the apex of the triangle prolonged into tail. It is found in salt water.

18. The *cyclidium* is frequently found in pure water, and has an oval, smooth, membranaceous, pellu-

cid body with a black margin. The tail is concealed under the edge, and comes out from it at every motion, but in such a manner as to project but little from the edge. There is also a kind of border to the hinder part.

19. The *tænia* appears like an oval pellucid membrane, something larger than the *monas hor*. The fore edge is thick and truncated; the hinder part acute, and terminating in a short tail. It whirls about in various directions with great velocity.

20. The *disfar* is a small orbicular animalcule, with a bent tail.

21. The *orbis* is round, and has a tail consisting of two long bristles.

22. The *luna* is likewise round, and has the fore-part hollowed into the form of a crescent.

XII. *Leucophra*.

An invisible, pellucid, and ciliated worm.

1. The *conflictor*, with moveable intestines, is perfectly spherical and semitransparent, of a yellow colour, the edges dark. It rolls from right to left, but seldom removes from the spot where it is first found. It is filled with a number of the most minute molecules, which move as if they were in a violent conflict; and in proportion to the number of these little combatants which are accumulated either on one side or other, the whole mass rolls either to the right or left. It then remains for a little time at rest, and the conflict ceases; but it soon becomes more violent, and the sphere moves the contrary way in a spiral line. When the water begins to fail, they assume an oblong, oval, or cylindric figure; the hinder part of some being compressed into a triangular shape, and the transparent part escaping as it were from the intestines, which continue to move with the same violence till the water fails, when the molecules shoot into a shapeless mass, which also soon vanishes, and the whole assumes the appearance of crystals of sal ammoniac.

2. The *mamilla* is of a dark colour, and filled with globular molecules; short hairs are curved inwards; and it occasionally projects and draws in a little white protuberance. It is pretty common in marshy water.

3. The *virginea* is a large, pear-shaped, greenish-coloured animalcule, filled with opaque molecules, and covered with short hairs; generally moving in a straight line. It is found in salt water.

4. The *viridis* is much smaller than the former, and cannot lengthen or shorten itself as it does. Sometimes it appears contracted in the middle, as if it were to be divided in two.

5. The *burfaria* is found in salt water, and is similar in many respects to the former. It is of a long oval shape, bulging in the middle, and filled with green molecules, every where ciliated except at the apex, which is truncated and shaped somewhat like a purse; the hairs are sometimes collected into little fascicles.

6. The *paftinaca* is globular, and covered as it were with a pellucid net; is found in fetid salt water.

7. The *aurea* is yellow, oval; has both ends equally obtuse; little hairs discovered with difficulty; and has in general a vehement rotatory motion.

8. The *parvula* is found in salt water; and is gelatinous

tissues and small, without any molecules. The fore-part is truncated, the hind-part brought nearly to a point, with a kind of oval hole on one side.

9. The *fcuta* is long, with sinuated angles, white, gelatinous, and granulated, changing its form considerably.

10. The *dilatata* appears like a gelatinous membrane, with a few grey molecules in the fore-part, and a great number in the hinder part. It is sometimes dilated into a triangular form with sinuated sides; at other times the shape is more irregular and oblong.

11. The *favillous* was found in December among the lesser lemnæ. It is of a green colour, oval, round, and opaque. It is supposed to be ciliated from its bright twinkling appearance, which probably arises from the motion it gives the water.

12. The *vesiculifera* is oval, very pellucid, with a defined dark edge and inside, containing some very bright bladders or vesicles. The middle frequently appears blue, and the vesicles appear as if set in a ground of that colour.

13. The *globulifera* was found in a ditch where the lemna minor grew. The body is round, very pellucid, without molecules, but with three little pellucid globules, and every where set with short hairs.

14. The *pustulata* is found in marshy waters; and is white, gelatinous, and somewhat granulated; the lower part truncated as if an oblique section were made in an egg near the bottom. It is covered with little erect shining hairs, and at the lower extremity a few bright pustules may be discovered.

15. The *turbinata* is found in thinking salt water; and is round, pellucid, somewhat of the shape of an acorn, with a pellucid globule at the lower end.

16. The *acuta* is found in salt water, and is gelatinous, thick, capable of assuming different shapes; having the apex bright, and the rest of the body filled with little spherules. Sometimes it draws itself up into an orbicular shape, at others one edge is sinuated.

17. The *natans* is oval, round, and has a black point at the edge.

18. The *candida* is found in salt water; and is membranaceous, flat, very white, with no visible intestines except two oval bodies not easily perceived. The whole edge is ciliated.

19. The *undulata* is oblong and oval, with a double row of little nodules.

20. The *lignata* is common in salt water in the months of November and December. It is oblong and subdepressed, with a black margin filled with little molecules, but more particularly distinguished by a curved line in the middle somewhat in the shape of the letter S; one end of which is sometimes bent into the form of a small spiral.

21. The *trigona* is found in marshes, but not commonly. It is a yellow triangular mass filled with unequal pellucid vesicles, one of which is much larger than the rest, and the edge surrounded with short fluctuating hairs.

22. The *fluida* is somewhat of a kidney shape, but ventricose.

23. The *flexa* is reniform and sinuated.

24. The *armilla* is round and annular.

25. The *cornuta* is of the shape of an inverted cone, opaque, and of a green colour. This requires to be

observed for some time before we can ascertain its character. The body is composed of molecular vesicles; the fore part is wide and truncated, with a little prominent horn or hook on both sides; the hind part being conical, every where ciliated, and the hairs exceedingly minute; those in the fore part are three times longer than the former, and move in a circular direction. The hinder part is pellucid, and sometimes terminates in two or three obtuse pellucid projections. At one time this animalcule will appear reniform and ciliated on the fore part; but at another time the hairs are concealed. It dissolves into molecular vesicles when the water evaporates.

26. The *heteroclita* appears to the naked eye like a white point; in the microscope as a cylindrical body, the fore part obtusely round, the middle rather drawn in; the lower part round, but much smaller than the upper part. It appears wholly ciliated through a large magnifier.

XIII. *Trichoda.*
An invisible, pellucid, hairy worm.

1. The *grandinella* is a very small pellucid globule, with the intestines scarce visible, the top of the surface furnished with several small bristles not easily discoverable, as the creature has a power of extending or drawing them back in an instant. It is found in pure water as well as in infusions of vegetables.

2. The *comata* is a pellucid globule filled with bright intestines, the fore part furnished with hairs, the hind part with a pellucid globule.

3. The *granata* resembles the two former; and has a darkish nucleus in the centre, with short hairs on the edge.

4. The *trochus* is somewhat of a pear-shape, and pellucid; each side of the fore part being distinguished by a little bunch of hairs.

5. The *gyrinus* is one of the smallest of this genus, and is found in salt water. It is smooth and free from hairs, except at the fore part, where there are a few.

6. The *fol* is small, globular, and crystalline; beset every where with diverging rays longer than the diameter of the body; the inside full of molecules. The body contracts and dilates, but the creature remains confined to the same spot. It was found with other animalcules in water which had been kept three weeks.

7. The *folaris* is orbicular, bright, and filled with globular intestines, frequently having in it a moveable substance of the shape of the letter S. It has hairs seldom exceeding 17 in number, set round the circumference, each of them nearly equal in length to the diameter of the animalcule.

8. The *lomba* is of a yellow colour, and full of clay-like molecules. It moves with such velocity as to elude the sight, and appears of various shapes, sometimes spherical, sometimes kidney-shaped, &c.

9. The *ovis* is composed of vesicular molecules; is of a spherical figure, smooth, pellucid, and a little notched in the fore part. The notched part is filled with long hairs, but there are none on the rest of the body.

10. The *urnula* is membranaceous, pellucid, somewhat in the form of a water pitcher, with the fore part hairy. It moves but slowly.

11. The *diota* is of a clay-colour, and filled with molecules;

Microscope molecules; the upper part cylindrical and truncated, the lower part spherical, the upper part of the mouth hairy at the edges.

12. The *horrida* is somewhat of a conical shape, the fore part rather broad and truncated, the lower part obtuse, and the whole covered with radiating bristles.

13. The *urinarium* is egg-shaped, with a short hairy beak.

14. The *lunellions* is smooth, pellucid, and shaped like a crescent.

15. The *trigonos* is of a triangular shape, a little convex on both sides, the fore part acute and ciliated, the hind part broader, and having the extremity as it were gnawed off.

16. The *situs* is round, not very pellucid, narrow in the fore part, and resembling an inverted club.

17. The *nigra* was found in salt water, and has an opaque body; but when at rest one side appears pellucid. When in violent motion, it seems entirely black.

18. The *pudex* is found in water where duckweed grows, chiefly in the month of December. It has a pouch above the hind part marked with black spots, depressed towards the top, a little folded, and somewhat convex on the under part. The apex is furnished with hairs, but they are seldom visible till the creature is in the agonies of death, when it extends and moves them vehemently, and attempting as it were to draw in the very last drop of water.

19. The *foccus* is membranaceous, the fore part rather conical, with three small hairy papillae projecting from the base.

20. The *fistuosa* is found in river water. It is oblong and depressed, with one margin hollow and hairy, and the lower end obtuse. It is of a yellow colour, and the hollow edge ciliated.

21. The *proserpe* is pellucid, the fore part formed into a kind of neck; one edge rising into a protuberance like a hump-back, the other edge convex.

22. The *proteus* is that which Mr Baker distinguishes by the same name, and of which an account is given under the article ANIMALCULE. It is found in the slimy matter adhering to the sides of vessels in which vegetables have been infused, or animal substances preserved. That described by Mr Adams was discovered in the slime produced from the water where small fishes, water-snails, &c. had been kept. The body resembled that of a snail, the shape being somewhat elliptical, but pointed at one end, while from the other proceeded a long, slender, and finely proportioned neck, of a size suitable to the rest of the animal.

23. The *versatilis* lives in the sea, and has some resemblance to the proteus; but the neck is shorter, the apex less spherical, and the hinder part of the trunk acute.

24. The *gibba* is pellucid; the upper part swelled out, with numerous molecules, and three large globules on the inside. The ends rather incline downwards; and when the water begins to fail, a few minute hairs may be discovered about the head and at the abdomen; the body then becomes striated longitudinally.

25. The *fustes* somewhat resembles a rolling-pin in shape; has both ends obtuse, and one shorter than the other. It can draw in the ends, and swell out the sides, so as to appear almost spherical.

26. The *pavus* is found in salt water; and is of a long cylindrical shape, filled with molecules, the fore part bright and clear, with a long opening near the top which tapers to a point, and is beset with hairs.

27. The *patula* is ventricose, rather inclining to an oval figure, with a small tube at the fore part, the upper part of which is hairy.

28. The *forcata* is oblong and rather broad, with three little horns on the fore part.

29. The *ferina* is found in the month of December in river-water. It is a beautiful animalcule, of a fox colour. It is of an oblong shape, the lower end somewhat larger than the other. It has a set of streaks running from one end to the other, and at the abdomen a double row of little eggs lying in a transverse direction.

30. The *scuda* is found in the infusion of hay and other vegetables. It is six times longer than broad, round, flexuous, of an equal size, the greater part filled with obscure molecules; the fore part rather empty, with an alimentary canal and lucid globules near the middle. The margin of the fore part is covered with short hairs.

31. The *aureola* is of a gold colour, pellucid, and filled with vehicles.

32. The *ignita* is of a fine purple colour, with something of a reddish cast, pellucid, splendid, with a number of globules of different sizes; the fore part small, the hinder part obtuse, with a very large opening which seems to run through the body.

33. The *prisma* is very small, and so transparent that it cannot easily be delineated. It is of a singular shape; the under part being convex, the upper compressed into a kind of keel, and the fore part small.

34. The *forceps* is found about the winter solstice in water covered with lemna. It is of a yellow colour, large, somewhat transparent, and filled with molecules, with a large opaque globule in the lower part. The fore part is divided into long lobes, one of which is falciform and acute, the other dilated and obliquely truncated. It can open, shut, or cross, those lobes at pleasure; and by the motion of them is appears to lock in the water.

35. The *forfex* is found in river water. It has the fore part formed into a kind of forceps, one of which is twice as long as the other, hooked and ciliated.

36. The *index* is found in salt water, and has the under part of the front of the margin hairy; the apex is formed by the fore part projecting like a finger on a direction post.

37. The *tricloda* is of a yellow colour, formed of two pellucid membranes striated longitudinally; the lower end obliquely truncated, and the two extremities bent in opposite directions.

38. The *cuvicula* has three corners; the fore part truncated and ciliated, the hind part acute and bent a little upwards. It has a crystalline appearance, and a kind of longitudinal keel runs down the middle.

39. The *flexilis* is of a flattened oval shape, the edge hairy, and hollowed out in such a manner as to form two unequal legs.

40. The *fulcata* is ovated and ventricose, the apex acute, with a furrow at the abdomen, and both sides of it ciliated.

41. The *mus* is found in pure water; and is smooth, five

Microscope five times broader than it is long, filled with darkish molecules. It has a bright neck, under the top of which are a few unequal hairs. It moves but languidly.

42. The *larbura* is round, somewhat linear, with both ends obtuse; the fore part narrower, forming as it were a kind of neck, under which is a row of fluctuating hairs. The trunk is full of grey molecules.

43. The *forcinus* is long, round, pellucid, and covered with very minute hairs, and has a great number of mucous vesicles about the body.

44. The *ursius* is long, round, every where ciliated on the upper part, and the under part likewise hairy as far as the middle.

45. The *angulus* is long, more convex than most of the genus, divided by a kind of articulation in the middle into two parts equal in breadth, but of different lengths; the apex has short waving hair.

46. The *linter* is found in an infusion of old grass. It is egg-shaped, oblong, with both extremities raised so that the bottom becomes convex, and the upper part depressed like a boat; it is of different shapes at different ages, and sometimes has a rotatory motion.

47. The *paxillus* is found in salt water; and is long, full of grey molecules; the fore part truncated and hairy, and rather smaller than the other.

48. The *vermicularis* is found in river water; and is pellucid in the fore part, with the hind part full of molecules.

49. The *melissa* is found in salt water, but very rarely. It is oblong, ciliated, with a globular apex, a dilatable neck, and a kind of peristaltic motion perceivable within it.

50. The *fimbriata* is subovated, the apex hairy, the hinder part obliquely truncated and serrated.

51. The *camulus* is found but rarely in vegetable infusions, and moves in a languid manner. The fore part is ventricose; the back divided by an incision in the middle into two tubercles; the lower part of the belly sinuated.

52. The *auger* is oblong, depressed, pellucid, and filled with molecules: the vertex is truncated, the fore-part forming a small beak with three feet underneath; beyond which, toward the hinder part, it is furnished with bristles.

53. The *papa* is roundish, pellucid, and consists of three parts. The head is broad, and appears to be hooded, the top being furnished with very small hairs; on the lower part of the head is a transparent vesicle, and over the breast from the base of the head hangs a production resembling the sheath of the feet in the papa of the gnat.

54. The *lunaris* is round and crystalline; the hinder part smaller than the other. The edge of the back and the part near the tail are bright and clear. It bends itself into the form of an arch.

55. The *lshuic* is arched and flattened with an hairy apex, and two little bristles proceeding from the tail.

56. The *rattus* is oblong, with a kind of keel; the fore part hairy, and a very long bristle proceeding from the hinder part.

57. The *tigris* resembles the former, but differs in the form of the tail, which consists of two bristles, and likewise in having a kind of incision in the body a little below the apex.

58. The *pavillus* is frequently found in marshes. It is cylindrical, pellucid, molecular, and capable of being folded up. It appears double; the interior part full of molecules, with an orbicular molecular appendage, which it can open and shut, and which forms the mouth. The external part is membranaceous, pellucid, dilated, and marked with transverse streaks; and it can protrude or draw in the orbicular membrane at pleasure. Some have four articulations in the tail, others five; and it has two pairs of bristles, one placed at the second joint, the other at the tail.

59. The *discus* has a considerable resemblance to a common nail; the fore part is round and hairy, the hinder part terminating in a sharp tail.

60. The *ovatus* is membranaceous, elliptical, full of molecules; the fore part hooded, the other round, and terminating in a tail as long as the body.

61. The *guttus* is found in river water. It is of a grey colour, flat, with seven large molecules and globules within it; the front obtuse, set with hairs; the hinder part terminating in a tail formed of very fine hairs.

62. The *naffulus* is found in the infusions of hay which have been kept for some months. It is roundish, egg-shaped, with a double margin drawn underneath it; the fore part narrow, and furnished with short hairs which continually play about; having a small tail underneath. It moves slowly, and is furnished with molecular intestines.

63. The *disfus* is found in river water. It is smooth, pellucid, having the fore part dilated into a semicircle, gradually decreasing in breadth towards the tail. The front is hairy, the hairs standing as rays from the semicircular edge: one of the edges is sometimes contracted.

64. The *delphinus* is found in hay that has been infused for some months. It is pellucid, smooth, and egg-shaped; the hinder part terminating in a tail about half the length of the body, dilated at the upper end, truncated, and always bent upwards. It moves sometimes on its belly and sometimes on its side.

65. The *clava*, or club *animalcula*, has the fore part thick, but the hinder part narrow; both extremities obtuse, pellucid, and replete with molecules; the hind part bent down towards the middle.

66. The *caudatus* is oblong, the fore part hairy, the hinder part rather acute, and filled with molecules and black vesicles.

67. The *felis* is large and curved, the fore part small, the hinder part gradually diminishing into a tail, the under part beset with hairs longitudinally.

68. The *fefus* is oblong, the fore part hairy, the hind part terminating in a very slender tail. It is smooth, pellucid, much longer than broad, and filled with yellow molecules; the fore part obtuse, the hinder part extremely slender and transparent, the upper side convex.

69. The *larva* is long, round, beset with hairs, and has the tail divided into two points.

70. The *linguauda* is cylindrical, the fore part truncated, and beset with hairs; the tail long, furnished with two bristles, and having two joints.

71. The *fax* has the circumference set with hairs, and a little solitary pedicle projecting from the body.

72. The *acpulaus* is sheathed within a cylindrical transparent

Microscope transparent bag, having a little pedicle bent back within the bag.

73. The *ingenita* is sheathed in a depressed bag, broadest at the base. The animalcule itself is funnel-shaped, with one or more hairs proceeding from each side of the mouth of the funnel. It can extend or contract itself within the bag, fixing its tail to the base, without touching the sides. It is found in salt water.

74. The *lunata* is sheathed in a cylindrical bag, with a pedicle passing through and projecting beyond it.

75. The *transfuga* is broad, the fore part hairy, the hinder part full of bristles; one side sinuated, and the other pointed.

76. The *ciliata* is ventricose, the hinder part covered with hair.

77. The *bulla* is membranaceous, the sides bent inwards; the fore and hind parts both covered with hairs.

78. The *pellonella* is somewhat thick in the middle, and pellucid, with a few molecules here and there; the sides obtuse, the fore part ciliated with very few hairs, the hinder part set with bristles.

79. The *crassum* has the hinder extremity filled with globules of various hues. It oscillates upon the edge, commonly advancing on its flat side, and continually drawing in water. It then gapes, and opens into a very acute angle, almost to the middle of the body; but this is done so instantaneously, that it can scarce be perceived.

80. The *cuspis* is oval, the fore part hairy, and the hinder part also furnished with some straight and curved hairs in two fascicles. Its body is flat, and filled with molecules; and in the fore part is an oblong empty space, into which we may sometimes see the water sucked in.

81. The *pulex* is egg-shaped, with an incision in the fore part; the front and base hairy.

82. The *lyncea* is nearly square, with a crooked beak and hairy mouth. It is membranaceous, and appears compressed, stretched out into a beak above, under which there is a little bundle of hairs; the lower edge bends in and out, and is surrounded with a few bristles. The intestines are beautiful, and a small bent tube goes from the mouth to them in the middle of the body. There is likewise another tube between the fore and hind edge, filled with blue liquor. The intestines and other tube are frequently in motion.

83. The *mola* is orbicular, the fore part notched; one side furnished with hairs, the hinder part with bristles.

84. The *restrata* is found in water where duckweed has been kept. It is depressed, capable of changing its shape, yellow, with long ciliated hairs; it has four feet tapering to a point, one of them longer than the rest. Both feet and hairs are within the margin. The shape of the body is generally triangular; the apex formed into an obtuse beak, which the creature sometimes draws in so that it appears quite round.

85. The *lagena* is round, ventricose, with a long neck, and the lower end set with bristles.

86. The *charon* was found in salt water. It is oval, and resembles a boat as well in its motion as shape: the upper part is hollowed, the under part furrowed.

N° 219.

and convex; the stern round, with several hairs proceeding from it.

87. The *cimex* is about the size of the lyncea, has an oval body, with a convex back, flat belly, and incision in the margin of the fore part, the edges of which incision appear to move. When this animalcule meets with any obstacles in swimming, it makes use of four bristles, which appear on the under side as feet.

88. The *cicada* differs but little from the cimex. It is oval, with an obscure margin, the fore part covered with hairs on the under side, and the hinder parts beardless.

XIV. Kerona.
An invisible worm with horns.

1. The *rugiolum* is found in river water. It has three rows of horns on the back, which occupy almost the whole of it.

2. The *lyncaster* is square, and its disc furnished with shining horns.

3. The *histrio* appears an oblong membrane, pellucid, with four or five black points in the fore part, which are continually changing their situation, thick set with small globules in the middle, among which four larger ones are sometimes perceived, which by Mr Adams are supposed to be eggs. In the middle of the hind part are some longitudinal strokes resembling bristles, which, however, do not seem to project beyond the body.

4. The *cypris* is found in water covered with lemna. It is somewhat of a pear shape, compressed, with a broad and blunt fore part; the front furnished with hairs or little vibrating points inserted under the edge, shorter in the fore part, partly extended straight, and partly bent down, having a retrograde motion.

5. The *bagttron* is orbicular, with the horns in the middle, the fore part membranaceous and ciliated, with several bristles at the hinder part.

6. The *bagtallum* differs from the preceding only in having the hinder part without any bristles.

7. The *patella* has an univalved shell, is orbicular, crystalline; the fore part somewhat notched; the fleshy body in the middle of the shell; with horns or hairs of different lengths jutting out beyond the shell, and acting instead of feet and oars, some of which are bent; and the superior ones constitute a double transverse row.

8. The *nassor* is oval and rather flat, with one edge bent, the opposite one ciliated, the front furnished with horns, and the hind part with bristles.

9. The *pulligilov* agrees in many respects with the *triclops pulex*; the upper part is pellucid, without any black molecules; the front trisected, the whole surface of the head covered with hair, and the fore part finuous.

10. The *mystilus* is a large animalcule; the fore and hind parts rounded, very pellucid and white, dark in the middle, with black intestines intermixed with a few pellucid vesicles; both extremities appearing as if composed of two thin plates. It has two small horns, with which it agitates the water so as to form a little whirlpool.

11. The *letur* is egg-shaped, compressed, pellucid,

5   and

Microscope and crowned with short waving hairs; the base terminated with bristles.

12. The *filarus* is an oval, smooth, animalcule, somewhat crooked and opaque, with a fascicle of vibrating hair on the fore part: it has a sharp tail furnished with unequal rows of moveable hairs, the back being also ciliated: the hairs produce a rotatory motion. The figure varies from oval to oblong, and the filaments of the conferva are often entangled in the tail.

13. The *colodium* is found in the infusion of vegetables. The body is broad and flat, both sides obtuse, filled with black molecules, and there is a black spot near the hinder part, where there are likewise a few short bristles.

14. The *pyftulans* is found in salt water. It is oval, convex; one edge of the hinder part sinuated, both ends set with hairs, and some horns on the fore part.

XV. *Himantopus*:
A pellucid, invisible, and cirrated worm.

1. The *acarus* is lively, conical, ventricose, full of black molecules, with a height and transparent fore part. The lower part of the apex has rows of long hairs on the under part set like rays. Four locks of long crooked hair or feet proceed from the belly, and it is continually moving these and other hairs in various directions.

2. The *ludio* is a lively diverting animalcule, smooth, pellucid, full of small points, the fore part clubbed and a little bent, the hinder part narrow; the base obliquely truncated, and terminating in a tail stretched out transversely. The top of the head and middle of the back are furnished with long and vibrating hairs; three moveable and flexible curls hang down from the side of the head at a distance from each other. When the creature is at rest, its tail is curled; but when in motion, it is drawn tight and extended upwards.

3. The *femen* is found, though seldom, in water where the herva grows. The cilia are longer than the hairs, and are continually vibrating: it has two moveable curls hanging on the side of the head.

4. The *volutator* is shaped like a crescent, and has some crystalline points; the convex part has a row of hairs lengeth towards the tail, and underneath are four feet. It is very lively, and often turns round with a swift circular motion.

5. The *larva* is long and cirrated in the middle; the body is depressed and long; the hinder parts acute, and generally curved, pellucid, and filled with granular molecules.

6. The *charon* is found in fen-water, but rarely. It is oval, pellucid, and membranous, with longitudinal furrows, and several feet diverging rows of hair below the middle, but none on the hinder part.

7. The *corona* is a membranous lamina, very thin, pellucid, crystalline, and semilunar: the edge of the base thick set with molecular interstices; the fore part furnished with a kind of mane; towards the hind part are three equal curved hairs or spines.

XVI. *Vorticella*:
A naked worm with rotatory cilia, capable of contracting and extending itself.

1. The *which* is visible to the naked eye, appearing

like a small green point; but the microscope discovers it to be exactly cylindrical, a little thicker at the fore part than the other, and obtuse at both ends. It appears to be totally destitute of limbs, notwithstanding which it keeps the water in continual motion; so that it probably has some invisible rotatory instrument. It moves sometimes circularly, sometimes in a straight line.

2. The *spherula* appears also like a point; but thro' the microscope as a globular mass of a dark green colour. It occasions a vehement motion in the water, probably by means of some short hairs with which it is furnished.

3. The *cincta* is of an irregular shape, sometimes assuming an oval figure, and appearing as if girt round with a transverse keel. It is invisible to the naked eye, ciliated on every side; the hairs all moveable, and longer on one side than the other.

4. The *lunifera* is found in salt water; has the fore-part obtuse, the base broad, and hollowed away like a crescent, with a short protuberance in the middle of the concave part: the fore part is ciliated.

5. The *lucifera* is found in salt water, and is ventricose, crammed with molecules; the fore part truncated, and both sides of it pellucid: there is a prominent papilla in the middle, which when the animalcule is at rest appears notched, the edge of the aperture being ciliated; the hairs are capable of moving in various directions.

6. The *curia* is cylindrical, truncated, opaque, and blackish coloured; the fore part ciliated.

7. The *spectorium* is found in October, with the lesser lenses, and is one of the most singular of the microscopic animalcules. When viewed sidewise, it is sometimes nearly cylindrical, only tapering a little towards the hinder part, and having a broad pellucid edge. Viewed from the top, it has sometimes a broad face or disc, furnished with radiating hairs, the under part contracted into a globular shape, of a dark green colour, and filled with small grains.

8. The *polymorpha* is visible to the naked eye, and appears like a green point moving with great agility; but when viewed through a microscope, it assumes such a variety of forms, that it is impossible to describe them. The body is granulous; and a series of pellucid points is sometimes to be observed.

9. The *multiformis* is found in salt water, and very much resembles the former.

10. The *nigra* is found in August in meadows covered with water. It may be seen with the naked eye, appearing like a black point swimming on the surface. Through the microscope it appears as a small conical body, obtuse and ventricose at one end and acute at the other. When the extremities are extended, two small white hooks become visible, by the assistance of which it moves in the water, and it probably has a rotatory organ: it moves continually in a vacillating manner on the top of the water.

11. The *cucullus* is likewise visible to the naked eye: it is of a dirty red colour, of a shape somewhat conical, and resembling a grenadier's cap.

12. The *utriculata* is green and ventricose; the belly capable of being lengthened or shortened; the fore part truncated, much in the shape of a common water bottle;

Microfcope bottle; the neck is sometimes very long, sometimes very short, and filled with green molecules.

13. The *cercaria* is met with in rivers, though very seldom, and in shape somewhat resembles the lower part of a boot. The apex of the upper part is truncated and ciliated, the heel pointed, and the foot round.

14. The *vulga* is as broad as long, and the apex truncated and ciliated; both angles of the base projecting outwards, one somewhat like a wart, the other like a finger. It is found in marshy waters.

15. The *papillaria* is likewise found in marshes where the conferva nitida grows. It is ventricole; the fore part truncated, with a papillary tail, and a beautiful papillary excrescence on the side.

16. The *faecalis* is thick, of an equal diameter every where, and full of molecules. The edge of the mouth is bent back; the hinder part is obtuse, sometimes notched and contracted, with cilia to be seen on both sides of the mouth.

17. The *cirrata* is found in ditch-water. It is ventricole, the aperture situated, and two tufts of hair on each side of the belly.

18. The *nasuta* is invisible to the naked eye, but the microfcope discovers it to be furnished with a rotatory organ encompassing the middle. It is pellucid, cylindrical, of an unequal size; the fore part truncated and ciliated, with a triangular prominence in the middle of the aperture; the hinder part is obtuse, with a point on each side of the middle of the body. When the water is nearly exhaled, two rotatory organs are obfervable; one on the fore part, and the other encompassing the middle of the body; the hairs of the latter being in violent motion. Other faetcules of moving hair are likewise to be obferved; and the quirk and various motions of this apparatus are very furprising.

19. The *stellina* is of an orbicular shape, with a molecular disc and ciliated margin.

20. The *diffina* is likewise orbicular, the edge ciliated, with a kind of handle on the under side.

21. The *feyplina* is bowl-shaped, crystalline, with an opaque fpherule in the middle.

22. The *albina* is cylindrical in the fore part, the hinder part tapering, and almoft ending in a point.

23. The *frittina* is empty and cylindrical, with a truncated apex.

24. The *truncatella* is of the larger kind of animalcules, with a crystalline body, full of black molecules, the skin perfectly smooth and colourless, the hinder extremity rounded, and the anterior part truncated: at this extremity there is a large opening that ferves for a mouth, which is thickly ciliated.

25. The *limacina* is cylindrical, truncated, and has two pair of cilia.

26. The *fraxinea* is mostly cylindrical, the hinder part rather tapering, and full of opaque molecules; transparent towards the upper end. Within the edge at the top are two small tubercles, from each side of which proceeds a pair of small hairs.

27. The *crategaria* is found in the month of April, both in the mud and on the tail of the monoculus quadricornis. They are generally heaped together in a fpherical form, and united to one common stalk. They are likewise often to be found without a pedicle, the body rather contracted, and the aperture circular, and fur-

rounded with a masked margin. It has two small arms; and with a powerful magnifier a violent rotatory motion may be obferved. Sometimes an individual will feparate from the community, and move in a kind of fpiral line for a little time, and then go back to the reft.

28. The *levmate* is not ciliated, nor has it any hairs upon it; the body is granulated, the fore part broad and truncated, the hinder part obtufe, and capable of being contracted or extended.

29. The *crateriformis* is a lively animalcule, pellucid, round, longer than it is broad, approaching fomewhat to a fquare figure, with convex fides: the head is fituated at the large end, the fkin fmooth, and fome traces of inteftines may be difcovered with difficulty. There is a confiderable opening furrounded by hair at the larger end, and the filaments compofing it are in continual motion. Two of thefe are fometimes feen joined together, and full of fmall fpherules. In this ftate they draw each other alternately different ways; the furface is fmooth, and the hairs invifible.

30. The *ramuloata* appears to the naked eye as a number of white points adhering to the fides of the glafs. When magnified, the fore part is narrower than the hind one; in the fide is a kind of incifion, and the hinder part is notched towards the middle. It excites a continual whirling motion in the water by means of a rotatory organ with which it is furnished.

31. The *corfpinofa* is a pellucid, gelatinous animalcule, of a greenifh colour, and furnished with fmall radii about the circumference; fo that it appears like a very fmall water hedge-hog.

32. The *ampulla* is contained in a transparent bottle-fhaped bag; the head divided into two lobes. It fometimes lies at the bottom of the bag, and fometimes nearly fills the whole of it.

33. The *folliculata* is gelatinous and cylindrical; and when moft extended, the bafe appears attenuated, and the apex truncated.

34. The *larva* is of a clay colour, fometimes ciliated, with a globular projection at times appearing to proceed from it.

35. The *funalata* has the fhape of an inverted cone, with an aperture in the figure of a crefcent; the lower part of the trunk notched, forming as it were two teeth; the tail biphyllous. Each of thefe is furrounded with a loofe bright fkin, the head being divided from the trunk by a deep incifion.

36. The *aurita* is cylindrical and ventricole, the aperture definite of hairs; both fides of it are furnished with rotatory cilia, and the tail is biphyllous.

37. The *arcuata* has fomething of a conical fhape; the mouth being divided into parts which are fet with fmall fpines; and a point projects from the tail.

38. The *felis* is molecular, pellucid, folding varioufly; the fore part truncated: round the margin are rows of hairs; but it has alfo ftiffer hairs or fpines continually vibrating, with which it draws in all animate and inanimate fubftances which it is able to manage.

39. The *lucinulata* is fhaped like an inverted cone, the aperture lobated, the tail fmall and furnished with two briftles. When fwimming, the rotatory organ may be difcovered. It moves fwiftly in an oblique direction.

40. The

Microscope 40. The *crossbsta* is of two kinds; viz. of a pale yellow and of a white colour. They move by fixing their tail to the place where they are, and then extending their body as much as possible; fixing the fore part to the place to which they intend to move, then drawing the hinder part to it, and so on. Sometimes they turn round about upon one of the points of their tail; at other times they spring forwards with a jerk. When at rest they open their mouths very wide.

41. The *tigate* has a convex body, filled with molecules, and of a dark colour; the hinder part somewhat broader than the forepart; the latter ciliated, and the tail formed of two very thin pellucid spines, which are somewhat curved, and much longer than the body.

42. The *rotaria* is the *wheel animal* described by Mr Baker; and of which an account is given under the article ANIMALCULE.

43. The *furcata* is commonly found in water, and has a cylindric body with a rotatory organ, consisting of a row of hairs at the apex; the tail is divided into two parts, turning a little inwards. When at rest it joins the segments of the tail, but opens them when in motion.

44. The *catulus* is commonly found in marshy waters. It is a little thick muscular animalcule, folding itself up; equally broad throughout, the body disfigured by longitudinal folds, winding in various directions. The anterior part is connected to the body by a little neck; and it occasionally shows a small rotatory organ. Its motion is rotatory, but in various directions.

45. The *conbula* is cylindrical, the aperture plain, with a short articulated tail divided into two parts.

46. The *pilo* has a large body, the apex of an equal thickness obtuse, with rotatory filaments: the tail is acute, with two pellucid spines in length about one-third part of the body, alternately separating from and approaching one another.

47. The *feminerus*. See the article POLYPE.

48. The *furcula*, when considerably magnified, appears like a circle surrounded with crowns or ciliated heads, and by small thin tails to a common centre, from whence they advance towards the circumference, where they turn very briskly, occasioning a kind of whirlpool, which brings its food. When one of them has been in motion for a time, it stops and another begins; sometimes two or three may be perceived in motion at once: they are frequently to be met with separate, with the tail sticking in the mud. The body contracts and dilates very much, so as sometimes to have the appearance of a cudgel, at others to assume almost a globular form.

49. The *stelligera* appears to the naked eye like a yellow globule adhering to the conserva; has a little flower or a heap of yellow eggs. When magnified, they are seen to consist of a congeries of animalcula constituting a sphere from a mouldy centre. They contract and extend their bodies either alone or in society, and excite a vortex in the water by means of a disk. When they quit the society and act singly, they may be observed to consist of a head, abdomen, and tail; the head being frequently drawn back into the abdomen so far that it cannot be seen, only exhibiting a broad kidney-shaped disk standing out. The abdo-

men is oblong, oval, and transparent; the tail sharp, Microscope twice as long as the abdomen, sometimes rough and annulated, or altogether smooth.

50. The *citrina* is found in stagnant water; the head full of molecules, round, every where of an equal size, and very transparent. Both sides of the orifice are ciliated, and each has a rotatory motion appearing sometimes without and sometimes within the edge of the mouth.

51. The *piriformis* is somewhat oval, with a very small retractile foot, which it can draw within itself.

52. The *tuberosa* has a broad upper part, the under part small, with two projections at the anterior end, furnished with a number of fibrillæ, which produce a current of water by their vibration, and thus collect food for the animal.

53. The *ringens* is pear-shaped, pellucid, the middle of the aperture convex, both sides ciliated, the pedicle four times shorter than the body. It can contract the orifice to an obtuse point.

54. The *inclinans* has a pendulous, pellucid, little head; the anterior part truncated, and occasionally contracting itself twice as short as the pedicle. It is shaped like a tobacco-pipe.

55. The *vaginata* is erect, of the shape of a truncated egg; the pedicle is contained in a sheath.

56. The *globularia* is frequent among the cyclogea quadricornis. It has a small spherical head, the aperture of the mouth ciliated, the pedicle four times longer than the body, which it contracts into a spiral form.

57. The *lunaris* has a small goblet-shaped head, the margin of the orifice protuberant, ciliated on both sides, with undulating hairs, and the pedicle eight or ten times the length of the body. The pedicle extends itself as often as the mouth is opened, but is twisted up spirally when it is shut; and this is frequently repeated in a short space.

58. The *convallaria* is the same with the *bell-animal* mentioned by Mr Baker. See the article ANIMALCULE.

59. The *anceos* has a simple pedicle; twists itself spirally; is extremely slender, with a kind of cup on its head; the margin white and round, and seemingly encompassed with a lucid ring; the head diminishing towards the base.

60. The *salpifera* is narrow at the base, open and truncated at the top; the margin seemingly surrounded with a ring; but, when the aperture is shut, the animalcule is of the shape of an egg, with a simple setaceous pedicle, considerably longer than the body, and commonly much bent back.

61. The *uncinaris* is visible to the naked eye: the head an inverted cone, convex when the mouth is shut, but truncated when it is open; with a protuberant edge; the pedicle simple, very long, thick, and whiter at the top than any where else; the apex twisted spirally.—When contracted, it appears to be annulated.

62. The *museta* inhabits that whitish substance which often entirely covers plants, wood, shells, &c. When this substance is examined by a microscope, it appears to be wholly composed of living animals of the polype kind. See POLYPE.

5 B a   63. The

Microscope   63. The *fasciculata* has a rotatory organ, which may sometimes be seen projecting beyond the aperture; there is a little head at the apex, and the pedicle is twisted and very slender. A congealed green mass which is often found swimming about in ditches is composed of myriads of these animals, which are not visible to the naked eye, and when magnified appear like a bundle of green flowers.

64. The *limon* resembles a citron; the apex is truncated, the base narrow, and a gaping cleft is observable, descending from the apex to one third of the body.

65. The *bellis* is of a yellow colour, and much resembles the flower of a daisy; is ciliated round the margin of the head, and moves in a rotatory manner.

66. The *gemella* has a long pedicle, constantly furnished with two small heads.

67. The *pyraria* }
68. The *anularia* } See the article POLYPE.
69. The *digitalis* }

71. The *polypina*, when viewed through a small magnifier, they appear like so many little trees: the upper part, or heads, are egg shaped, the top truncated, the lower part filled with intestines; the branches thick set with little knobs.

72. The *racemosa* is only distinguished from the vorticella socialis by always adhering to the sides of the vessel in which it is placed. By the microscope, we discover a long pedicle sticking to the sides of the vessels, from which proceed an innumerable quantity of crystalline pellucid pearls; which, together with the stalk, are variously agitated in the water. Sometimes they move separately; sometimes they are drawn down to the root, and in a moment expanded again.

XVII. *Brachionus*:
A contractile worm, covered with a shell, and furnished with rotatory cilia.

1. The *bractea* has an oblong, pellucid shell, capable of altering its figure. The apex is truncated, with six small teeth on the edge of it, twelve longitudinal streaks down the back, the base obtuse and smooth. The teeth are occasionally protruded or retracted; and there are two small spines or horns on the other side of the shell. The animal itself is of a yellow colour, crystalline, and muscular; now and then putting out from the apex two or three little bundles of playing hairs, the two lateral ones shorter than that in the middle: on the under side we may observe a forked deglutatory muscle, and two rigid points when the apex is drawn in. It is found in sea-water.

2. The *squamula* has an univalve orbicular shell, a truncated apex, four teeth, smooth base, and no tail.

3. The *pala* is of a yellow colour; univalved, with an oblong excavated shell; four long teeth at the apex; a smooth base.

4. The *bipalium* is univalved, the shell oblong and inflected, ten teeth at the apex, the base smooth, and a spurious tail.

5. The *patina* is extremely bright and splendid, has a large body, a crystalline and nearly circular shell, without either footion or teeth, only towards the apex it falls in in as to form a smooth notch. A double glittering organ, with ciliated edges, projects from the apex; both of them of a conical figure, and standing

as it were upon a pellucid substance, which is divided into two lobes, between which and the rotatory organ there is a silver-coloured crenulated membrane. Two small claws may likewise be discovered near the mouth.

6. The *clypeatus* is univalved, the shell oblong, apex notched, the tail naked, and base smooth.

7. The *lunularis* is univalved; the shell extending considerably beyond the body; the base divided into three small horns, with two hairs at the end of the tail.

8. The *patella* is found in marshy water in the winter-time. It is univalve, the shell oval, plain, crystalline, with the anterior part terminating in two acute points on both sides, though the intervening space is commonly filled up with the head of the animal. By these points it fastens itself, and whirls about the body erect. The rotatory cilia are perceived with great difficulty.

9. The *bractea* is univalved, the shell somewhat orbicular, apex lunated, base smooth, and the tail furnished with two spines.

10. The *plicatilis* is univalved, with an oblong shell, the apex hairy, and base notched.

11. The *ovalis* is bivalved; the shell flattened, apex notched, a hollow part at the base, the tail formed of two tufts of hair.

12. The *tripos* is bivalved, the apex of the shell beardless, three horns at the base, and double tail. It fixes itself to objects by the filaments of the tail.

13. The *dentatus* is bivalved, with an arched shell; the apex and base are both toothed, and the tail formed of two spines.

14. The *marronotus* is bivalved, somewhat of a square form; the base and apex pointed; the tail consisting of two spines.

15. The *urcinatus* is one of the smallest bivalved animalcules; the apex and anterior part round, the hinder part straight, terminating in a point, furnished with a hook on the fore part, a small rotatory organ, a long tail composed of joints, and divided at the end into two bristles. It can open its shell both at the fore and hind part.

16. The *circulus* is larger than the preceding; ventricose, somewhat transparent, the head conical, with a bundle of hairs on both sides; and it has likewise a rotatory organ.

17. The *passus* has a cylindric shell, with two long pendulous locks of hair proceeding from the front, the tail consisting of a single bristle.

18. The *quadratus* has a quadrangular shell, with two small teeth at the apex, two horns proceeding from the hair, and no tail.

19. The *impositus* has a quadrangular shell, a smooth undivided apex; obtuse base; notched margin; and flexuous tail.

20. The *secularis*. See POLYPE.

21. The *brachionus Bakeri* has a ventricose shell, four teeth at the apex, two horns at the base, and a long tail terminating in two short points. The horns are frequently extended; and the circular end of each is furnished with a tuft of little hairs, which sometimes move in a vibratory manner, at other times have a rotatory motion. Mr Muller has also discovered in this creature two small feelers and a tongue.

22. The *patulus* has a ventricose shell, with eight teeth.

Plate CCCII.

Microscopic Objects.

Fig. 38.

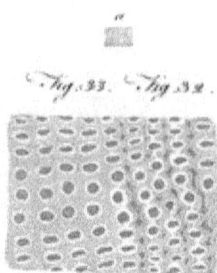

Fig. 33. Fig. 32.

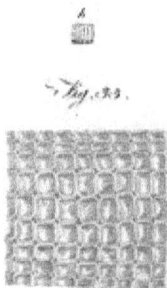

Fig. 35.

Fig. 36.

Fig. 37.

Fig. 39.

Fig. 40.

Fig. 41.

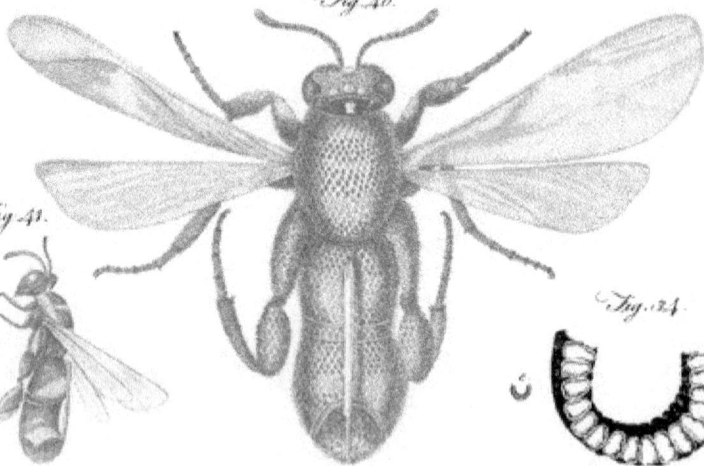

Fig. 34.

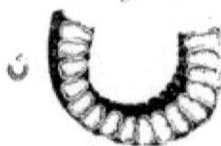

J. A. Bell Fecit. W.J. Sculptor fecit.

teeth at the apex; the base humated, or hollowed into the form of a crescent, and furnished with four horns; the tail short, with two small points at the end.

These are the different kinds of animalcules which have yet been discovered. To what is said of them in general under the article ANIMALCULE, we shall here add the following observations from Mr Adams.—

" How many kinds of these invisibles there may be (says he) is yet unknown; as they are diffused of all sizes, from those which are barely invisible to the naked eye, to such as resist the force of the microscope as the fixed stars do that of the telescope, and with the greatest powers hitherto invented appear only as so many moving points. The smallest living creatures our instruments can show, are those which inhabit the waters; for though animalcula equally minute may fly in the air, or creep upon the earth, it is scarce possible to get a view of these; but as water is transparent, by confining the creatures within it we can easily observe them by applying a drop of it to the glasses.

" Animalcules in general are observed to move in all directions with equal ease and rapidity, sometimes obliquely, sometimes straight forward; sometimes moving in a circular direction, or rolling upon one another, running backwards and forwards through the whole extent of the drop, as if diverting themselves; at other times greedily attacking the little parcels of matter they meet with. Notwithstanding their extreme minuteness, they know how to avoid obstacles, or to prevent any interference with one another in their motions; sometimes they will suddenly change the direction in which they move, and take an opposite one; and, by inclining the glass on which the drop of water is, as it can be made to move in any direction, so the animalcules appear to move as easily against the stream as with it. When the water begins to evaporate, they flock towards the place where the fluid is, and show a great anxiety and uncommon agitation of the organs with which they draw in the water. These motions grow languid as the water fails, and at last cease altogether, without a possibility of renewal if they be left dry for a short time. They sustain a great degree of cold as well as insects, and will perish in much the same degree of heat that destroys insects. Some animalcules are produced in water at the freezing point, and some insects live in snow.—By mixing the least drop of urine with the water in which they swim, they instantly fall into convulsions and die.

" The same rule seems to hold good in those minute creatures, which is observable in the larger animals, viz. that the larger kinds are less numerous than such as are smaller, while the smallest of all are found in such multitudes, that there seem to be myriads for one of the others. They increase in size, like other animals, from their birth until they have attained their full growth; and when deprived of proper nourishment, they in like manner grow thin and perish."

The modes of propagation among these animalcules are various, and the observation of them is extremely curious. Some multiply by a transverse division, as is observed under the article ANIMALCULE: and it is re-

markable, that though in general they avoid one another, there, it is not uncommon, when one is nearly divided, to see another push itself upon the small neck which joins the two bodies in order to accelerate the separation.—Others, when about to multiply, fix themselves to the bottom of the water; then becoming first oblong, and afterwards round, turn rapidly as on a centre, but perpetually varying the direction of their rotatory motion. In a little time, two lines forming a cross are perceived; after which the spherule divides into four, which grow, and are again divided as before. A third kind multiply by a longitudinal division, which in some begins in the fore part, in others in the hind part; and from others a small fragment detaches itself, which in a short time assumes the shape of the parent animalcule. Lastly, others propagate in the same manner as the more perfect animals.

In our observations under the article ANIMALCULE, we suggested some doubts whether all those minute bodies which go under the name of animalcules really do enjoy animal life; or whether they are not in many cases to be accounted only inanimate and exceedingly minute points of matter actuated by the internal motion of the fluid. This has also been the opinion of others: but to all hypotheses of this kind Mr Adams makes the following reply. " From what has been said, it clearly appears, that these motions are not purely mechanical, but are produced by an internal spontaneous principle; and that they must therefore be placed among the class of living animals, for they possess the strongest marks and the most decided characters of animation; and, consequently, that there is no foundation for the supposition of a chaotic and neutral kingdom, which can only have derived its origin from a very transient and superficial view of these animalcules.—It may also be farther observed, that as we see that the motions of the limbs, &c. of the larger animals, are produced by the mechanical construction of the body, and the action of the soul thereon, and are forced by the ocular demonstration which arises from anatomical dissection to acknowledge this mechanism which is adapted to produce the various motions necessary to the animal; and as, when we have recourse to the microscope, we find those pieces which had appeared to the naked eye as the primary mechanical causes of particular motions, to consist themselves of lesser parts, which are the causes of motion, extension, &c. in the larger; when the structure therefore can be traced no farther by the eye, or by the glasses; we have no right to conclude that the parts which are invisible are not equally the subject of mechanism; for this would be only to affirm, in other words, that a thing may exist because we see and feel it, and have no existence when it is not the object of our senses.—The same train of reasoning may be applied to microscopic insects and animalcula; we see them move; but because the muscles and members which occasion these motions are invisible, shall we infer that they have not muscles, with organs appropriated to the motion of the whole and its parts? To say that they exist not because we cannot perceive them, would not be a rational conclusion. Our senses are indeed given us that we may comprehend some effects; but then we have also a mind, with reason, bestowed upon

Microscope upon us, that, from the things which we do perceive with our senses, we may deduce the nature of those causes and effects which are imperceptible to the corporeal eye."

Leaving these speculations however, we shall now proceed to give a particular

Explanation of the figures of the various animals, with their parts, &c. represented in the plates.

Plate CCCXII.

Fig. 32. 33. represent the eggs of the phalæna neustria, as they are taken from the tree to which they adhere, and magnified by the microscope. The strong ground-work visible in many places shows the gum by which they are fastened together; and this connection is strengthened by a very tenacious substance interposed between the eggs, and filling up the vacant spaces. Fig. 34. shows a vertical section of the eggs, exhibiting their oval shape.— Fig. 35. is an horizontal section through the middle of the egg. These eggs make a beautiful appearance through the microscope. The small figures *a, b, c,* represent the objects in their natural state, without being magnified.

Fig. 36. shows the larva of the *musca chamæleon,* an aquatic insect. When viewed by the naked eye, it appears (as here represented) to be composed of twelve annular divisions, separating it into an head, thorax, and abdomen; but it is not easy to distinguish the two last parts from each other, as the intestines lie equally both in the thorax and abdomen. The tail is furnished with a fine crown or circle of hair *b,* disposed in the form of a ring, and by this means it is supported on the surface of the water, the head and body hanging down towards the bottom, in which posture it will sometimes remain for a considerable time without any motion.— When it has a mind to sink to the bottom, it closes the hairs of the ring, as in fig. 37. Thus an hollow space is formed, including a small bubble of air; by enlarging or diminishing which, it can rise or sink in the water at pleasure. When the bubble escapes, the insect can replace it from the pulmonary tubes, and sometimes considerable quantities of air may be seen to escape from the tail of the worm into the common atmosphere; which operation may easily be observed when the worm is placed in a glass of water, and affords an entertaining spectacle. The snout of this insect is divided into three parts, of which that in the middle is immoveable; the other two, which grow from the sides of the middle one, are moveable, and vibrate like the tongues of lizards or serpents. In these lateral parts lies most of the creature's strength; for it walks upon them when out of the water, appearing to walk on its mouth, and to use it as the parrot does its beak to assist it in climbing.

The larva is shown fig. 38. as it appears through a microscope. It grows narrower towards the head, is largest about that part which we may call the thorax, converges all along the abdomen, and terminates at length in a sharp tail surrounded with hairs, as has already been mentioned. The twelve annular divisions are now extremely visible, and are masked by numbers in the plate. The skin appears somewhat hard, and resembling shagreen, being thick set with grains pretty equally distributed. It has nine holes, or spiracula, probably for the purpose of breathing, on each side;

but it has none of these on the tail division *a,* nor any easily visible on the third from the head. In the latter, indeed, it has some very small holes concealed under the skin, near the place where the embryo wings of the future fly are laid. " It is remarkable (says Mr Adams) that caterpillars, in general, have two rings without these spiracula, perhaps because they change into flies with four wings, whereas this worm produces a fly with only two." The skin of the larva is adorned with oblong black furrows, spots of a light colour, and orbicular rings, from which these generally springs a hair; but only those hairs which grow on the insect's sides are represented in the figure. There are also some larger hairs here and there, as at *c c.* The difference of colours, however, in this worm arises only from the quantity of grains in the same space; for where they are in very great numbers, the furrows are darker, and paler where they are less plentiful.

The head *d* is divided into three parts, and covered with a skin which has hardly any discernible grains.— The eyes are rather protuberant, and lie near the snout; on which last are two small horns at *i i.* *i i* is crooked, and ends in a sharp point as at *f.* The legs are placed near the snout between the sinuses in which the eyes are fixed. Each of these legs consists of three joints, the outermost of which is covered with stiff hairs like bristles *g g.* From the next joint there springs a horny bone *h h,* used by the insect as a kind of thumb; the joint is also composed of a black substance of an intermediate hardness between bone and horn; and the third joint is of the same nature. In order to distinguish these parts, those that form the upper sides of the mouth and eyes must be separated by means of a small knife; after which, by the assistance of the microscope, we may perceive that the leg is articulated by some particular ligaments, with the portion of the insect's mouth which answers to the lower jaw in the human frame. We may then also discern the muscles which serve to move the legs, and drive them up into a cavity that lies between the bones and those parts of the mouth which are near the horns *i i.* The insect walks upon these legs, not only in the water, but on the land also. It likewise makes use of them in swimming, keeping its tail on the surface contiguous to the air, and hanging downward with the rest of the body in the water. In this situation, the only perceptible motion it has is in its legs, which it moves in a most elegant manner, from whence it is reasonable to conclude, that the most of this creature's strength lies in its legs, as we have already observed.

The snout of this larva is black and hard; the back part quite solid, and somewhat of a globular form; the front *f* sharp and hollow. Three membranaceous divisions may be perceived on the back part; by means of which, and the muscles contained in the snout, the creature can contract or expand it at pleasure.

The extremity of the tail is surrounded with thirty hairs, and the sides adorned with others that are smaller; and here and there the large hairs branch out into smaller ones, which may be reckoned single hairs. All these have their roots in the outer skin, which in this place is covered with rough grains, as may be observed by cutting it off and holding it against the light upon

Microscope upon a slip of glass. Thus also we find, that at the extremities of the hairs there are grains like those on the skin; and in the middle of the tail there is a small opening, within which are minute holes, by which the insect takes in and lets out the air it breathes. These hairs, however, are seldom disposed in such a regular order as is represented in fig. 38. unless when the insect floats with the body in the water, and the tail with its hairs a little lower than the surface, in which case they are disposed exactly in the order delineated in the plate. The least motion of the tail downward produces a concavity in the water; and it then assumes the figure of a wine-glass, wide at the top and narrow at the bottom. The tail answers the double purpose of swimming and breathing, and through it the insect receives what is the principle of life and motion to all animals. By means of these hairs also it can stop its motion when swimming, and remain suspended quietly without motion for any length of time. Its motions in swimming are very beautiful, especially when it advances with its whole body floating on the surface of the water after filling itself with air by the tail.— To set out, it first bends the body to the right or left, and then contracts it in the form of the letter S, and again stretches it out in a straight line; by thus contracting and then extending the body alternately, it moves on the surface of the water. It is very quiet, and is not disturbed by handling.

These creatures are commonly found in shallow standing waters in the beginning of June; but some years much more plentifully than others. They crawl on the grass and other plants which grow in such waters, and are often met with in ditches floating on the surface of the water by means of their tail, the head and thorax at the same time hanging down; and in this posture they turn over the clay and dirt with their feet and feet in search of food, which is commonly a viscous matter met with in small ponds and ditches. It is very harmless, though its appearance would seem to indicate the contrary. It is most easily killed for dissection by spirit of turpentine.

Fig. 39. shows in its natural size a beautiful insect, described by Linnæus under the name of Linnaeus and which appears to be a kind of intermediate genus between a sphex and a wasp. The antennæ are black and cylindrical, increasing in thickness towards the extremity to the joint nearest the head is yellow; the head and thorax are black, encompassed with a yellow line, and furnished with a cross line of the same colour near the head. The scutellum is yellow, the abdomen black, with two yellow bands, and a deep spot of the same colour on each side between the bands. A deep polished groove extends down the back from the thorax to the anus, into which the sting turns and is deposited, leaving the anus very circular; a yellow line runs on each side of the sting. The anus and whole body, when viewed with a small magnifier, appear punctuated; but when these points are seen through a large magnifier, they appear hexagonal. Fig. 40. shows the insect very much magnified. Fig. 41. gives a side view of it magnified in a smaller degree.

Fig. 42. shows an insect lately discovered by Mr John Adams of Edmonton, as it appeared to be an ion. It was first seen by some labouring people who were there at the time, by whom it was conjectured to be a louse with unusually long horns, a mite, &c. Mr Adams hearing the debate, procured the insect; and having viewed it through a microscope, it presented the appearance exhibited in fig. 42. The insect seems to be quite distinct from the phalangium cancroides of Linnæus. The latter has been described by several authors, but none of their descriptions agree with this. The abdomen of this insect is more extended, the claws larger, and much more obtuse; the body of the other being nearly orbicular, the claws slender, and almost terminating in a point, more transparent, and of a paler colour. Mr Markham has one in his possession not to be distinguished from that represented in fig. 42, excepting only that it wants the break or dent in the claws, which is so conspicuous in this. He found that insect firmly fixed by its claws to the thigh of a large fly, which he caught on a flower in Essex in the first week of August, and from which he could not disengage it without great difficulty, and tearing off the leg of the fly. This was done upon a piece of writing paper; and he was surprised to see the little creature spring forward a quarter of an inch, and again seize the thigh with its claws, so that he had great difficulty in disengaging it. The natural size of this creature, which Mr Adams calls the bladder-insect, is exhibited at a.

Fig. 43. shows the insect named by M. de Geer Physapus, on account of the bladders at its feet, (Thrips physapus, Lin.) This insect is to be found in great plenty upon the flowers of dandelion, &c. in the spring and summer. It has four wings, two upper and two under ones (represented fig. 44.) but the two undermost are not to be perceived without great difficulty. They are very long; and fixed to the upper part of the breast, lying horizontally. Both of them are rather pointed towards the edges, and have a strong nerve running round them, which is set with a hair fringe tufted at the extremity. The colour of these wings is whitish; the body of the insect is black; the head small, with two large reticular eyes. The antennæ are of an equal size throughout, and divided into six oval pieces, which are articulated together.— The extremities of the feet are furnished with a membranaceous and flexible bladder, which it can throw out or draw in at pleasure. It presses this bladder against the substances on which it walks, and thus seems to fix itself to them; the bladder sometimes appears concave towards the bottom, the concavity diminishing as it is less pressed. The insect is represented of its natural size at b.

Fig. 45. represents the thrips fasciata of Linnæus, remarkable for very bright and elegantly disposed colours, though few in number. The head, probolcis, and thorax, are black: the thorax ornamented with yellow spots; the middle one large, and occupying almost one-third of the posterior part; the other two are on each side, and triangular. The scutellum has two yellow oblong spots, pointed at each end. The ground of the elytra is a bright yellow, spotted and striped with black. The nerves are yellow; and there is a brilliant triangular spot of orange, which unites the cruftaceous and membrana-

Microscope count parts; the latter are brown, and clouded. It is found on the elm-tree in June. It is represented of its natural size at *e*.

Fig. 46. shows the *Chrysomela asparagi* of Linnæus, so called from the larva of the insect feeding upon that plant. It is a common insect, and very beautiful. It is of an oblong figure, with black antennæ, composed of many joints, nearly oval. The head is a deep and bright blue; the thorax red and cylindrical; the elytra are blue, with a yellow margin, and having three spots of the same colour on each; one at the base, of an oblong form, and two united with the margin; the legs are black; but the under side of the belly is of the same blue colour with the elytra and head. This little animal, when viewed by the naked eye, scarcely appears to deserve any notice; but when examined by the microscope, is one of the most pleasing opaque objects we have. It is found in June on the asparagus after it has run to seed; and it is shown of its natural size at *d*. De Geer says that it is very scarce in Sweden.

Fig. 47. shows an insect of a shape so remarkable, that naturalists have been at a loss to determine the genus to which it belongs. In the Fauna Suecica, Linnæus makes it an *unidica*; but in the last edition of the *Systema Naturæ*, it is ranged as a *meloe*, under the title of the *Meloe monoceros*; though of this also there seems to be some doubt. The true figure of it can only be discovered by a very good microscope. The head is black, and appears to be hid or buried under the thorax, which projects forward like a horn; the antennæ are composed of many joints, and are of a dirty yellow colour, as well as the feet; the hinder part of the thorax is reddish, the fore-part black. The elytra are yellow, with a black longitudinal line down the suture; there is a band of the same colour near the apex, and also a black point near the base, the whole animal being curiously covered with hair. The natural size of it is shown at *s*. It was found in May. Geoffroy says that it lives upon umbelliferous plants.

Fig. 48—53. exhibit the anatomy of the cossus caterpillar, which lives on the willow. The egg from which it proceeds is attached to the trunk of the tree by a kind of viscous juice, which soon becomes so hard that the rain cannot dissolve it. The egg itself is very small and spheroidal, and, when examined by the microscope, appears to have broad waving furrows running through the whole length of it, which are again crossed by close streaks, giving it the appearance of a wicker basket. It is not exactly known what time they are hatched; but as the small caterpillars appear in September, it is probable that the eggs are hatched some time in August. When small, they are generally met with under the bark of the tree to which the eggs were affixed; and an aqueous moisture, oozing from the hole through which they got under the bark, is frequently, though not always, a direction for finding them. These caterpillars change their colour but very little, being nearly the same when young as when old. Like many others, they are capable of spinning as soon as they come from the egg. They also change their skin several times; but as it is almost impossible to rear them under a glass, so it is very difficult to know exactly how often this moulting takes place.— Mr Adams conjectures that it is more frequently than

N° 239.

the generality of caterpillars do, some having been observed to change more than nine times.

The cossus generally falls for some days previous to the moulting; during which time the fleshy and other interior parts of the head are detached from the old skull, and retire as it were within the neck. The new coverings soon grow on, but are at first very soft.— When the new skin and the other parts are formed, the old skin is to be opened, and all the members withdrawn from it; an operation naturally difficult, but which must be rendered more so from the soft and weak state of the creature at that time. It is always much larger after each change.

From Mr Lyonet's experiments, it appears, that the cossus generally passes at least two winters, if not three, before it attains the pupa state. At the approach of winter, it forms a little case, the inside of which is lined with silk, and the outside covered with wood ground like very fine saw-dust. During the whole season it neither moves nor eats.

This caterpillar, at its first appearance, is not above one-twelfth of an inch long; but at last attains the length of two, and sometimes of three inches. In the month of May it prepares for the pupa state; the first care being to find a hole in the tree sufficient to allow the moth to issue forth; and if this cannot be found, it makes one equal in size to the future pupa. It then begins to form of wood a case or cone; uniting the bits, which are very thin, together by silk, into the form of an ellipsoid, the outside being formed of small bits of wood joined together in all directions; taking care, however, that the pointed end of the case may always be opposite to the mouth of the hole: having finished the outside of the case, it lines the inside with a silken tapestry of a close texture in all its parts, except the pointed end, where the texture is looser, in order to facilitate its escape at the proper time. The caterpillar then places itself in such a posture, that the head may always lie towards the opening of the hole in the tree or pointed end of its case. Thus it remains at rest for some time; the colour of the skin first becomes pale, and afterwards brown; the interior parts of the head are detached from the skull; the legs withdraw themselves from the exterior case; the body shortens; the posterior part grows small, while the anterior part swells so much, that the skin at last bursts; and, by a variety of motions, is pushed down to the tail; and thus the pupa is exhibited, in which the parts of the future moth may be easily traced.— The covering of the pupa, though at first soft, humid, and white; soon dries and hardens, and becomes of a dark purple colour: the posterior part is moveable; but not the fore-part, which contains the rudiments of the head, legs, and wings. The fore-part of the pupa is furnished with two horns, one above and the other under the eyes. It has also several rows of points on its back. It remains for some weeks in the case; after which the moth begins to agitate itself, and the points are them of essential service, by acting as a fulcrum, upon which it may rest in its endeavours to proceed forward, and not slip back by its efforts for that purpose.

The moth generally continues its endeavours to open the case for a quarter of an hour; after which, by redoubled efforts, it enlarges the hole, and presses forward

Fig. 34.

Fig. 46.

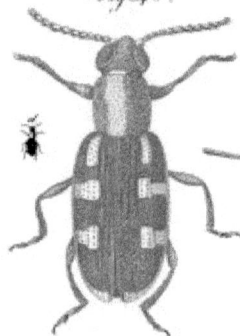

Fig. 42.

Fig. 43.

Fig. 44.

Fig. 49. Fig. 50.

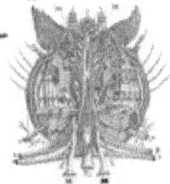

Fig. 45.

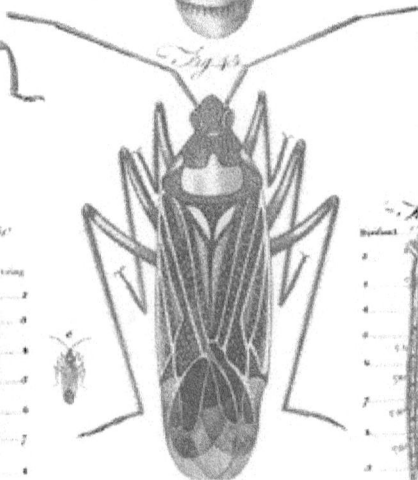

Fig. 51. Fig. 50.

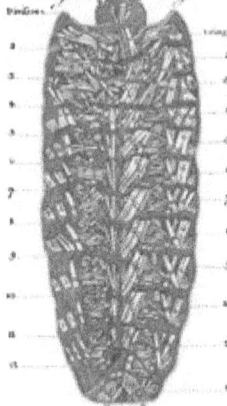

Fig. 48. Fig. 49.

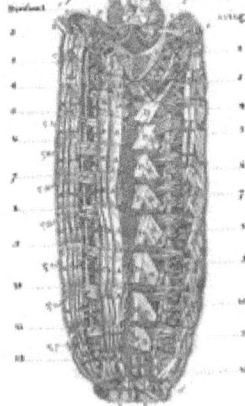

Fig. 47.

A. Bell Pine St. Sculptor, fecit.

Microscope ward until it arrives at the edge, where it makes a full stop, lest by advancing further it should fall to the ground. After having in this manner reposed itself for some time, it begins to disengage itself entirely; and having rested for some hours with its head upwards, it becomes fit for action. Mr Martham says, that it generally pushes one third of the case out of the hole before it halts.

The body of the caterpillar is divided into twelve rings, marked 1, 2, 3, &c. as represented in fig. 48. 49. 50. 51. each of which is distinguished from that which precedes, and that which follows, by a kind of neck or hollow; and, by forming boundaries to the rings, we make twelve other divisions, likewise expressed in the figures; but to the first of these the word ring is affixed, and to the second, division. To facilitate the description of this animal, M. Lyonet supposed a line to pass down through the middle of the back, which he called the superior line, because it marked the most elevated part of the back of the caterpillar; and another, passing from the head down the belly to the tail, he called the inferior line.

All caterpillars have a small organ, resembling an elliptic spot, on the right and left of each ring, excepting the second, third, and last; and by these we are furnished with a further subdivision of this caterpillar, viz. by lines passing through the spiracula, the one on the right side, the other on the left of the caterpillar. These four lines, which divide the caterpillar longitudinally into four equal parts, mark each the place under the skin which is occupied by a considerable viscus. Under the superior line lies the heart, or rather thread of hearts; over the inferior line, the spinal marrow; and the two tracheal arteries follow the course of the lateral lines. At equal distances from the superior and two lateral lines, we may suppose four intermediate lines. The two between the superior and lateral lines are called the intermediate superior; the two others opposite to them, and between the lateral and inferior lines, are called the intermediate inferior.

Fig. 48. 49. show the muscles of the caterpillar, arranged with the most wonderful symmetry and order, especially when taken off by equal strata on both sides, which exhibits an astonishing and exact form and correspondence in them. The figures show the muscles of two different caterpillars opened at the belly, and supposed to be joined together at the superior line. The muscles of the back are marked by capitals; the gastric muscles by Roman letters; the lateral ones by Greek characters. Those marked ⁰ are called, by M. Lyonet, dividing muscles, on account of their situation.

The caterpillars are prepared for dissection by being emptied, and the muscles, nerves, &c. freed from the fat in the manner formerly directed; after which the following observations were made.

The muscle A in the first ring is double; the anterior one being thick at top, and being apparently divided into different muscles on the upper side, but without any appearance of this kind on the under side. One insertion is at the skin of the neck towards the head; the other is a little above; and that of the second muscle A is a little below the first spiraculum, near which they are fixed to the skin.

The muscle marked - is long and slender, fixed by its anterior extremity under the gastric muscles a and b of the first ring, to the circumflex scale of the base of the lower lip. It communicates with the muscle c of the second ring, after having passed under some of the arteries, and introduced itself below the muscle b.

The muscle b is so tender, that it is scarce possible to open the belly of the caterpillar without breaking it. It is sometimes double, and sometimes triple. Anteriorly it is fixed to the posterior edge of the side of the parietal scale, the lower suture being at the middle of the ring near the inferior line.

There are three muscles marked c; the first affixed at one extremity near the lower edge of the upper part of the parietal scale; the other end divides itself into three or four tails, fixed to the skin of the caterpillar under the muscle b. The anterior part of the second is fixed near the first; the anterior part of the third a little under the first and second, at the skin of the neck under the muscle A. These two last passing over the cavity of the first pair of limbs, are fixed by several tails to the edge opposite to this cavity. In this subject there are two muscles marked b, but sometimes three is only one anteriorly; they are fixed to the lower edge of the parietal scale, the other ends being inserted in the first fold of the skin of the neck on the belly-side. Fig. 50. best represents the muscles b and c; as in that figure they do not appear injured by any unnatural connection.

In the second and four following rings we discern two large dorsal muscles A and B. In the 7th, 5th, and 11th rings are three, A, B, and C; in the 11th are four, A, B, C, and D; and in the anterior part of the 12th ring are five, A, B, C, D, and E. All these ranges of muscles, however, as well as the gastric muscles a, b, c, d, appear at first sight only as a single muscle, running nearly the whole length of the caterpillar; but when this is detached from the animal, it is found to consist of so many distinct muscles, each consisting only of the length of one of the rings, their extremities being fixed to the division of each ring, excepting the middle muscle a, which, at the 6th, 7th, 8th, and 9th rings, has its insertions rather beyond the division. Each row of muscles appears as one, because they are closely connected at top by some of the fibres which pass from one ring to the other.

The muscles A, which are 12 in number, gradually diminish in breadth to the lower part of the left ring; at the 8th and three following divisions they communicate with the muscles B, and at the 11th with D. In the lower part of the last ring, A is much broader than it was in the preceding ring; one extremity of it is contracted, and communicates with B; the lower insertion being at the membrane I, which is the exterior skin of the fecal bag. The muscles A and B, on the lower part of the last ring, cannot be seen until a large muscle is removed, which on one side is fixed to the subdivision of the ring and on the other to the fecal bag.

The right muscles D, which are also 12 in number, begin at the second ring, and grow larger from thence to the seventh. They are afterwards narrower from thence to the 12th; the deficiency is width be-

ing

Microscope supplied by the six muscles C, which accompany it from the 7th to the subdivision of the 12th ring. The muscles B and C communicate laterally with the 8th, 11th, and 12th divisions. C is wanting at the subdivision of the 12th; its place being here supplied by B, which becomes broader at this part.

The first of the three floating muscles V originates at the first ring, from whence it introduces itself under N, where it is fixed, and then subdivides, and hides itself under other parts. The second begins at the second division, being fixed to the anterior extremity B of the second ring; from thence directing itself towards the stomach; and, after communicating with the case of the *corpus cæcum*, it divides, and spreads into eight muscles which run along the belly. The third begins at the third division, originating partly at the skin, and partly at the junction of the muscles B of the second and third ring. It directs itself obliquely towards the belly, meeting it near the third spiraculum; and branching from thence, it forms the oblique muscles of some of the anterior viscera.

The thin, long, muscle l, which is at the subdivision of the last ring, and covers the anterior insertion of the muscle (m) where the ring terminates, is single. It begins at one extremity of the muscle (c); at the fore-part of the ring runs along the subdivision round the belly of the caterpillar, and finishes, on the other side, at the extremity of a similar muscle C.

Fig. 49. shows the dorsal muscles of the coffin. To view which in an advantageous manner, we must use the following mode of preparation.

1. All the dorsal muscles, 35 in number, must be taken out, as well as the seven lateral ones already described.

2. All the straight muscles of the belly must be taken away, as well as the muscular roots (r), and the ends of the gastric muscles (c), which are at the third and fourth divisions.

3. At the second division the muscle s must be removed; only the extremities being left to show where it was inserted.

The parts being thus prepared, we begin at the third ring; where there are found four dorsal muscles C, D, E, and F. The first one C, is inserted at the third division, under the muscles s and s, where it communicates by means of some fibres with the muscle f of the second ring; proceeding from thence obliquely towards the intermediate superior line, and is fixed at the fourth division. As soon as C is retrenched, the muscle D is seen; which grows wider from the anterior extremity: it lies in a contrary direction to the muscle C, and is inserted into the third and fourth divisions. The muscle E lies in the same direction as the muscle C, but not so obliquely: the lower insertion is at the fourth division; the other at the third, immediately under C. The muscle F is nearly parallel to D which joins it; the first insertion is visible; but the other is hid under the muscles E and G at the fourth division.

In the eight following rings, there are only two dorsal muscles; and of these D is the only one that is completely seen. It is very large, and diminishes gradually in breadth from one ring to the other, till it comes to the last, sending off branches in some

places.—E is one of the strait muscles of the back; and is inserted under the dividing muscles s, at the divisions of its own ring.

On the anterior part of the 12th ring there are three dorsal muscles, D, E, and F. D is similar to that of the preceding ring, marked also D, only that it is no more than half the length; terminating at the subdivision of its own ring. E is of the same length, and differs from the muscle E of the preceding ring only in its direction. F is parallel to E, and shorter than it; its anterior end does not reach the twelfth division.

On the posterior part there is only one dorsal muscle, fastened by some short ones to the subdivision of the last ring, traversing the muscles s; and being fixed there as if designed to strengthen them, and to vary their direction — s is a single muscle, of which the anterior insertion is visible, the other end being fixed to the bottom of the foot of the last leg; its use is to move the foot. The anterior part of the muscle s branches into three or four heads, which cross the superior line obliquely, and are fixed to the skin a little above it. The other end is fastened to the membrane T.

Fig. 50. and 51. show the muscles of the caterpillar when it is opened at the back. The preparation for this view is to disengage the fat and other extraneous matter, as before directed.

The first ring has only two gastric muscles (z) and (d); the former is broad, and has three or four little tails; the first fixture is at the base of the lower lip, from whence it descends obliquely, and is fixed between the inferior and lateral line. The small muscle (d) is fastened on one side to the first spiraculum; on the other, a little lower, to the intermediate inferior and lateral line; and seems to be an antagonist to the muscle P, which opens the spiracula. The posterior fixture of s is under the muscle C, near the skin of the neck: s is fixed a little on the other side of C, at the middle of the ring.

In the second ring there are three gastric muscles, g, b, and b: z and b are fixed at the folds which terminate the ring; but only the anterior part of s is fixed there. The muscle b is triple, and in one of the divisions separated into two parts, that marked s comes nearer the inferior line, and is fixed a little beyond the middle of the ring, where the corresponding muscle of the opposite side is forked to receive it.

In the third ring, the muscle b, which was triple in the foregoing ring, is only double here, that part which is nearest the inferior line being broadest; it has three tails, of which only two are visible in the figure. It is exactly similar to that of the preceding ring; and is crossed in the same manner by the muscle from the opposite side of the ring.

Throughout the eight following rings, the muscle f which runs through them all is very broad and strong. The anterior part of it is fixed at the intermediate inferior line, on the fold of the first division of the ring: the other part is fixed beyond the lower division; with this difference, that at the 10th and 11th rings it is fixed at the last fold of its ring; whereas, in the others it passes over that ring, and is inserted into the skin of the following one. In all

their,

Microscope these, the first extremity of the muscle *g* is fastened to the fold which separates the ring from the preceding one, and is parallel to *f*, and placed at the side of it. The six first muscles marked *g*, are forked; that of the fourth ring being more so than the rest, nor does it unite till near its anterior insertion. The longest tail lays hold of the following, and is inserted near the inferior line; the other inserts itself near the same line, at about the middle of its own ring. The two last do not branch out; but terminate at the divisions, without reaching the following ring. The muscle *b*, placed at the side of *f*, has nearly the same direction, and finishes at the folds of the ring.

The anterior part of the 12th ring has only one gastric muscle, marked *e*: it is placed on the intermediate inferior line; and is inserted at the folds of the upper division, and at the subdivision of this ring. The lower part has a larger muscle marked *c*, with several divisions; one placed under *b*, with one extremity fixed near the lateral line, at the subdivision of its ring; the other to the fecal bag, a little lower than the muscle *b*.

In fig. 51. all the gastric muscles described in fig. 50. disappear, as well as those lateral and dorsal ones of which the letters are not to be found in this figure.

In the first ring are the gastric muscles, *e*, *f*, *g*, which are best seen here: the first is narrow and long, passing under and crossing *f*; one of its insertions is at the lower line, the other at the lateral, between the spiraculum and neck; *f* is short, broad, and nearly straight, placed along the intermediate line; but between it and the lateral it passes under *e*, and is fixed to the fold of the skin which goes from the one bag to the other; the lower insertion is near the second division. There are sometimes three muscles of those marked *g*, and sometimes four: the lower parts of them are fixed about the middle of the ring, and the anterior parts at the fold of the skin near the neck. The muscles *i* and *b* are fixed to the same fold; the other end of *b* being fixed under the muscle *h*, near the spiraculum. Above the upper end of *f*, a muscular body, *g*, may be seen. It is formed by the separation of two floating muscles.

The second ring has six gastric muscles, *b*, *l*, *m*, *n*, *o*, *p*. The first is a large oblique muscle, with three or four divisions placed at the anterior part of the ring: the head is fixed between the inferior line and its intermediate one, at the fold of the second division; from whence it crosses the inferior line and its corresponding muscle, terminating to the right and left of the line. *l* is a narrow muscle, whose head is fixed to the fold of the second division; the tail of it lying under *n*, and fastened to the edge of the skin that forms the cavity for the leg. The two muscles marked *m* have the same obliquity, and are placed the one on the other: the head is inserted in the skin under the muscle *l*, and communicates by a number of fibres with the tail of the muscle *v*; the other end is fixed to the intermediate inferior line at the fold of the third division. The large and broad muscle *n*, covers the lower edge of the cavity of the limb, and the extremity of the tail of *l*. It is fixed first at the skin, near the intermediate line, from whence it goes

in a perpendicular direction towards *n*, and introduces Microscope itself under *o* and *m*, where it is fixed. The muscle *o* is narrow and bent, and covers the edge of the cavity of the leg for a little way; one end terminating there, and the other finishing at the third division near *n*. That marked *p* is also bent: it runs near the anterior edge of the cavity of the leg: one end meets the head of *o*, the other end terminates at a raised fold near the inferior line. There is a triangular muscle on the side of the lateral muscle *o*, similar to that marked *g* in the following ring: in this figure it is entirely concealed by the muscle *n*.]

The third ring has no muscle similar to *m*; that marked *k* differs only from that of the second ring in being crossed by the opposite muscle. Those marked *l*, *n*, *o*, *f*, are similar to those of the preceding one. The muscle *g* is triangular; the base is fastened to the tail fold of the ring; on the lower side it is fixed to the muscle *n*, the top to the skin at the edge of the cavity for the leg.

The eight following rings have the gastric muscles, *i*, *k*, *l*, and *n*. The muscle *i* is quite straight, and placed at some distance from the inferior line: it is broad at the fourth ring, but diminishes gradually in breadth to the 11th. In the fourth it is united; but divides into two heads, which divaricate in the following rings. In the six next rings these heads are fixed nearly at the same place with *o* and *f*; and in the other two it terminates at the fold of the ring. The anterior insertion of the first and last is at the fold where the ring begins: that of the six others is somewhat lower under the place where the muscle *i* terminates. The lower part of the oblique muscle *b* is inserted in the skin near *i*; the upper part at the intermediate inferior muscle upon the fold which separates the following ring, but is wanting in the 11th. The muscle *l* is large, and co-operates with M: in the opening and shutting the spiraculum, one of its fixtures is near the intermediate inferior line, at about the same height as *i*. The tail terminates a little below the spiraculum.

The twelfth ring has only the single gastric muscle *d*, which is a bundle of six, seven, or eight muscles: the first fixture of these is at the subdivision of the ring near the inferior line: one or two cross this, and at the same time the similar muscles of the opposite side. Their fixture is at the bottom of the foot; and their office is to assist the muscle *a* in bringing back the foot, and to loosen the claw from what it lays hold of. One of the insertions of this muscle *a* is observed in this figure near *d*, the other near the subdivision of the ring.

Fig. 52. and 53. show the organization of the head of the collet, though in a very imperfect manner, as M. Lyonet found it necessary to employ twenty figures to explain it fully. The head is represented as it appears when separated from the fat, and disengaged from the neck. H H are the two palpi. The truncated muscles D belong to the lower lip, and assist in moving it. K shows the two ganglions of the neck united. I I are the two reels which assist in spinning the silk. L, the oesophagus. M, the two dissolving vessels. The Hebrew characters ×ɔɔ show the continuation of the four or five arteries. In fig. 52. the ten abductor muscles of the jaw are represented by SS, TT, VV, and Z. Four occipital muscles are seen in fig. 53. under *e e* and *f f*. At *a a* is represented a nerve of the first pair belonging to the ganglion of the neck; *b* is a branch of this nerve.

Fig. 54. exhibits the nerves as seen from the under part; but excepting in two or three nerves, which may be easily distinguished, only one of each pair is drawn, in order to avoid confusion. The nerves of the real ganglion of the neck are marked by capital letters, those of the ganglion (a) of the head by Roman letters; the nerves of the small ganglion by Greek characters. Those of the frontal ganglion, except one, by numbers.

The muscles of the coffus have neither the colour nor form of those of larger animals. In their natural state they are soft, and of the confistence of a jelly. Their colour is a greyish blue, which, with the silver-coloured appearance of the pulmonary vessels, form a glorious spectacle. After the caterpillar has been looked for some time in spirit of wine, they lofe their elasticity and transparency, becoming firm, opaque, and white, and the air-vessels totally disappear. The number of muscles in a caterpillar is very great. The greatest part of the head is composed of them, and there is a vast number about the œsophagus, intestines, &c. the skin is, as it were, lined by different beds of them, placed the one under the other, and ranged with great symmetry. M. Lyonet has been able to distinguish 228 in the head, 1647 in the body, and 2066 in the intestinal tube, making in all 4041.

At first fight the muscles might be taken for tendons, as being of the same colour, and having nearly the same hue. They are generally flat, and of an equal fize throughout; the middle feldom differing either in colour or hue from either of the extremities. If they are separated, however, by means of very fine needles, in a drop of fame fluid, we find them composed not only of fibres, membranes, and air-vessels, but likewise of nerves; and, from the drops of oil that may be feen floating on the fluid, they appear also to be furnished with many unctuous particles. Their ends are fixed to the skin, but the rest of the muscle is generally free and floating. Several of them branch out confiderably; and the branches fometimes extend fo far, that it is not eafy to difcover whether they are diftinct and feparate muscles or parts of another. They are moderately ftrong; and thofe which have been foaked in fpirit of wine, when examined by the microfcope, are found to be covered with a membrane which may be feparated from them; and they appear then to confift of feveral parallel bands lying longitudinally along the mufcle, which, when divided by means of fine needles, appear to be compofed of ftill fmaller bundles of fibres lying in the fame direction; which, when examined by a powerful magnifier, and in a favourable light, appear twifted like a fmall cord. The mufcular fibres of the fpiders, which are much larger than thofe of the caterpillar, confift of two different fubftances, one foft and the other hard; the latter being twifted round the former fpirally, and thus giving it the twifted appearance juft mentioned.

There is nothing in the caterpillar fimilar to the brain in man. We find indeed in the head of this insect a part from which all the nerves feem to proceed; but this part is entirely unprotected, and fo fmall, that it does not occupy one tifth part of the head: the furface is fmooth, and has neither lobes nor any anfractuofity like the human brain. But if we call this a brain in the caterpillar, we muft fay that it has eleven; for there are twelve other fuch parts following each other in a ftraight line, all of them of the fame fubftance with that in the head, and nearly of the fame fize; and from them, as well as from that in the head, the nerves are diftributed thro' the body.

The fpinal marrow in the coffus goes along the belly; is very fmall, forking out at intervals, nearly of the fame thicknefs throughout, except at the ganglions, and is not inclofed in any cafe. It is by no means fo tender as in man; but has a great degree of tenacity, and does not break without a confiderable degree of tenfion. The fubftance of the ganglions differs from that of the fpinal marrow, as no vessels can be difcovered in the latter; but the former are full of very delicate ones. There are 94 principal nerves, which divide into innumerable ramifications.

The coffus has two large tracheal arteries, creeping under the fkin clofe to the fpiracles; one at the right and the other at the left fide of the infect, each of them communicating with the air by means of nine fpiracula. They are nearly as long as the whole caterpillar; beginning at the firft fpiraculum, and extending fomewhat farther than the laft; fome branches alfo extending quite to the extremity of the body. Round each fpiraculum the trachea pufhes forth a great number of branches, which are again divided into fmaller ones, and thefe further fubdivide and fpread through the whole body of the caterpillar. The tracheal artery, with all its numerous ramifications, are open elaftic vessels, which may be preffed clofe together, or drawn out confiderably, but return immediately to their ufual fize when the tenfion ceafes. They are naturally of a filver colour, and make a beautiful appearance. This vessel, with its principal branches, is compofed of three coats, which may be feparated from one another. The outmoft is a thick membrane furnifhed with a great variety of fibres, which defcribe a vaft number of circles round it, communicating with each other by numerous fhoots. The fecond is very thin and tranfparent, without any particular vessel being diftinguifhable in it. The third is compofed of fcaly threads, generally of a fpiral form; and fo near each other as fcarcely to leave any interval. They are curioufly united with the membrane which occupies the intervals; and form a tube which is always open, notwithftanding the flexure of the vessel. There are alfo many other peculiarities in its ftructure. The principal tracheal vessels divide into 1326 different branches.

The heart of the coffus is very different from that of larger animals, being almoft as long as the animal itfelf. It lies immediately under the fkin at the top of the back, entering the head, and terminating near the mouth. Towards the laft rings of the body it is large and capacious, diminifhing very much as it approaches the head, from the fourth to the twelfth divifion. On both fides, at each divifion, it has an appendage, which partly

Plate CCCVI.

Microscopic Objects.
Fig. 55.

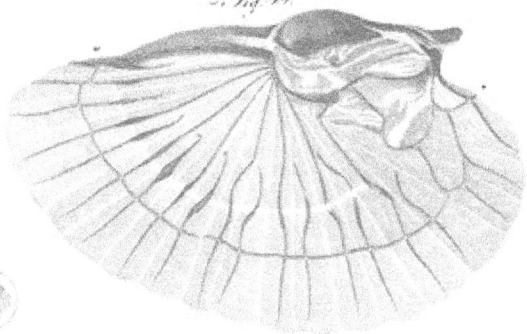

Fig. 58.

Fig. 57.

Fig. 59.

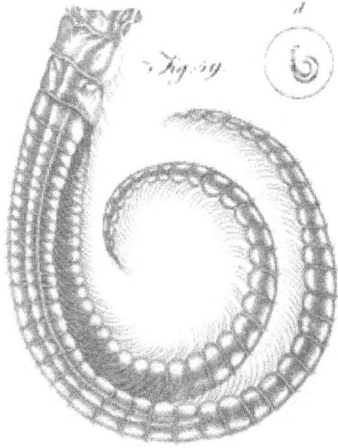

Fig. 56.

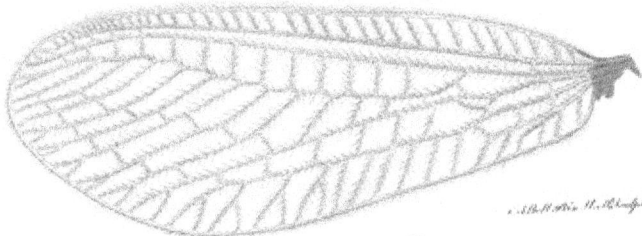

Fig. 62.

Fig. 65.

Fig. 64.

b

l

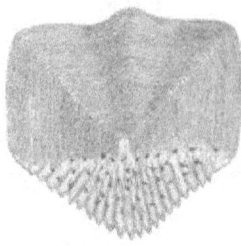

h

Fig. 61.

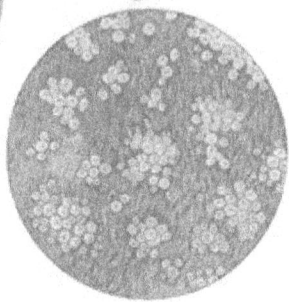

Fig. 63.

Fig. 66.

l

g

Fig. 60.

f

a   c   b   d   b   c   a

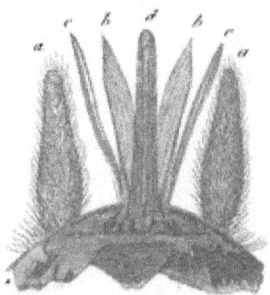

C   B

D   E

J. Sell, Phia. Philadelphia, feci.

Microscope partly covers the muscles of the back, but which, growing narrower as it approaches the lateral line, it forms a number of irregular lozenge-shaped bodies.— This tube, however, seems to perform none of the functions of the heart in larger animals, as we find no vessel opening into it which answers either to the aorta or vena cava. It is called the heart, because it is generally filled with a kind of lymph, which naturalists have supposed to be the blood of the caterpillar; and because in all caterpillars which have a transparent skin, we may perceive alternate regular contractions and dilatations along the superior line, beginning at the eleventh ring, and proceeding from ring to ring, down the fourth; whence this vessel is thought to be a string or row of hearts. There are two white oblong bodies which join the heart near the eighth division; and these have been called *auxiliar bodies*, from their having somewhat of the shape of a kidney.

The most considerable part of the whole caterpillar with regard to bulk is the corpus crassum. It is the first and only substance that is seen on opening it. It forms a kind of sheath which envelopes and covers all the entrails, and, introducing itself into the head, enters all the muscles of the body, filling the greatest part of the empty spaces in the caterpillar. It very much resembles the configuration of the human brain, and is of a milk-white colour.

The oesophagus descends from the bottom of the mouth to about the fourth division. The fore-part, which is in the head, is fleshy, narrow, and fixed by different muscles to the crustaceous parts of it; the lower part, which passes into the body, is wider, and forms a kind of membranaceous bag, covered with very small muscles; near the stomach it is narrower, and, as it were, confined by a strong nerve fixed to it at distant intervals. The ventricle begins a little above the fourth division, where the oesophagus ends, and finishes at the tenth. It is about seven times as long as broad; and the anterior part, which is broadest, is generally folded. These folds diminish with the bulk as it approaches the intestines; the surface is covered with a great number of aerial vessels, and opens into a tube, which M. Lyonet calls the large intestine.— There are three of these large tubes, each of which differs so much from the rest, as to require a particular name to distinguish it from them.

The two vessels from which the coffus spins its silk are often above three inches long, and are distinguished into three parts; the anterior, intermediate, and posterior. It has likewise two other vessels, which are supposed to prepare and contain the liquor for dissolving the wood on which it feeds.

Plate CCCIV.
Fig. 55. shews the wing of an earwig magnified; *a* represents it of the natural size. The wings of this insect are so artificially folded up under short cases, that few people imagine they have any. Indeed, they very rarely make use of their wings. The cases under which they are concealed are not more than a sixth part of the size of one wing, though a small part of the wing may be discovered, on a careful inspection, projecting from under them. The upper part of the wing is crustaceous and opaque, but the under part is beautifully transparent. In putting up their wings, they first fold back the parts A B, and then shut up the ribs like a fan; the strong muscles used for this purpose being seen at the upper part of the figure. Some of

the ribs are extended from the centre to the outer edge; others only from the edge about half way; but they are all assisted by a kind of bone, at a small but equal distance from the edge; the whole evidently contrived to strengthen the wing, and facilitate its various motions. The insect often differs very little in appearance in its three different states. De Geer asserts, that the female hatches eggs like a hen, and broods over her young ones as a hen does.

Fig. 56. represents a wing of the *Hemerobius perla* magnified. It is an insect which seldom lives more than two or three days.—The wings are nearly of a length, and exactly similar to one another. They are composed of fine delicate nerves, regularly and elegantly disposed as in the figure, beautifully adorned with hairs, and lightly tinged with green. The body is of a fine green colour; and its eyes appear like two burnished beads of gold, whence it has obtained the name of *golden eye*. This insect lays its eggs on the leaves of the plum or the rose tree; the eggs are of a white colour, and each of them fixed to a little pedicle or foot-stalk, by which means they stand off a little from the leaf, appearing like the fructification of some of the mosses. The larva proceeding from these eggs resembles that of the coccinella or lady-cow, but is much more handsome. Like that, it feeds upon aphides or pucerons, sucking their blood, and forming itself a case with their dried bodies; in which it changes into the pupa state, from whence they afterwards emerge in the form of a fly.

Fig. E, F, I, represent the dust of a moth's wing magnified. This is of different figures in different moths. The natural size of these small plumes is represented at H.

Fig. 57. shows a part of the cornea of the libellula magnified. In some positions of the light, the sides of the hexagons appear of a fine gold colour, and divided by three parallel lines. The natural size of the part magnified is shown at *b*.

Fig. 58. shows the part *c* of a lobster's cornea magnified.

Fig. 59. shows one of the arms or horns of the lepas anatifera, or barnacle, magnified; its natural size being represented at *d*. Each horn consists of several joints, and each joint is furnished on the concave side of the arm with long hairs. When viewed in the microscope, the arms appear rather opaque; but they may be rendered transparent, and become a most beautiful object, by extracting out of the interior cavity a bundle of longitudinal fibres, which runs the whole length of the arm. Mr Needham thinks that the motion and use of these arms may illustrate the nature of the rotatory motion in the wheel-animal. In the midst of the arms is an hollow trunk, consisting of a jointed hairy tube, which incloses a long round tongue that can be pushed occasionally out of the tube or sheath, and retracted occasionally. The mouth of the animal consists of six laminæ, which go off with a bend, indented like a saw on the convex edge, and by their circular disposition are so ranged, that the teeth, in the alternate elevation and depression of each plate, act against whatever comes between them. The plates are placed together in such a manner, that to the naked eye they form an aperture not much unlike the mouth of a contracted purse.

Fig. 60. shows the apparatus of the *Talanus* or Gad-fly,

Plate CCCV.

Microscope fly, by which it pierces the skin of horses and oxen, in order to suck their blood. The whole is contained in a fleshy case, not expressed in the figure. The feelers a a are of a spongy texture and grey colour, covered with short hairs. They are united to the head by a small joint of the same substance. They defend the other parts of the apparatus, being laid upon it side by side whenever the animal flings, and thus preserve it from external injury. The wound is made by the two lancets b b and B, which are of a delicate structure, but very sharp, formed like the dissecting knife of an anatomist, growing gradually thicker to the back.— The two instruments c c and C, appear as if intended to enlarge the wound, by irritating the parts round it; for which they are jagged or toothed. They may also serve, from their hard and horny texture, to defend the tube c E, which is of a softer nature, and tubular, to admit the blood, and convey it to the stomach. This part is totally inclosed in a tube of D, which entirely covers it. These parts are drawn separately at B, C, D, E. De Geer observes, that only the females suck the blood of animals; and Reaumur informs us, that having made one, that had sucked its fill, disgorge itself, the blood is threw up appeared to him to be more than the whole body of the insect could have contained. The natural size of this apparatus is shown at f.

Fig. 61. shows a bit of the skin of a lump-fish (Cyclopterus) magnified. When a good specimen of this can be procured, it forms a most beautiful object. The tubercles exhibited in the figure probably secrete an unctuous juice.

Fig. 62. shows the scale of a sea-perch found on the English coast; the natural size is exhibited at b.

Fig. 63. is the scale of an haddock magnified; its natural size as within the circle.

Fig. 64. the scale of a parrot fish from the West Indies magnified; f the natural size of it.

Fig. 65. the scale of a kind of perch in the West Indies magnified; d the natural size of the scale.

Fig. 66. part of the skin of a sole fish, as viewed through an opaque microscope; the magnified part, in its real size, shown at h.

The scales of fishes afford a great variety of beautiful objects for the microscope. Some are long; others round, square, &c. varying considerably not only in different fishes, but even in different parts of the same fish. Leeuwenhoeck supposed them to consist of an infinite number of small scales or strata, of which those next to the body of the fish are the largest. When viewed by the microscope, we find some of them ornamented with a prodigious number of concentric flutings, too near each other, and too fine to be easily enumerated. These flutings are frequently traversed by others diverging from the centre of the scale, and generally proceeding from thence in a straight line to the circumference.

For a more full information concerning these and other microscopical objects, the reader may consult Mr Adams's *Essays on the Microscope*, who has made the most valuable collection that has yet appeared on the subject. See also the articles Animalcule, Crystallization, Polype, Plants, and Wood, in the present Work.

MIDAS (fab. hist.), a famous king of Phrygia, who having received Bacchus with great magnificence,

that god, out of gratitude, offered to grant him whatever he should ask. Midas desired that every thing he touched should be changed into gold. Bacchus consented; and Midas, with extreme pleasure, everywhere found the effects of his touch. But he had soon reason to repent of his folly: for wanting to eat and drink, the aliments no sooner entered his mouth than they were changed into gold. This obliged him to have recourse to Bacchus again, to beseech him to restore him to his former state; on which the god ordered him to bathe in the river Pactolus, which from thence forward had golden sands. Some time after, being chosen judge between Pan and Apollo, he gave another instance of his folly and bad taste, in preferring Pan's music to Apollo's; on which the latter being enraged, gave him a pair of asses ears. This Midas attempted to conceal from the knowledge of his subjects: but one of his servants saw the length of his ears, and being unable to keep the secret, yet afraid to reveal it from apprehension of the king's resentment, he opened a hole in the earth, and after he had whispered there that Midas had the ears of an ass, he covered the place as before, as if he had buried his words in the ground. On that place, where the poets mention, grew a number of reeds, which when agitated by the wind uttered the same sound that had been buried beneath, and published to the world that Midas had the ears of an ass. Some explain the fable of the ears of Midas, by the supposition that he kept a number of informers and spies, who were continually employed in gathering every seditious word that might drop from the mouths of his subjects. Midas, according to Strabo, died of drinking bull's hot blood. This he did, as Plutarch mentions, to free himself from the numerous ill dreams which continually tormented him. Midas, according to some, was son of Cybele. He built a town which he called Ancyra.

MIDAS, *Ear-shell*, the smooth ovato-oblong buccinum, with an oblong and very narrow mouth. It consists of six volutions, but the lower one alone makes up almost the whole shell.

MID-HEAVEN, the point of the ecliptic that culminates, or in which it cuts the meridian.

MIDDLEBURG, one of the Friendly Islands in the South Sea. This island was first discovered by Tasman, a Dutch navigator, in January 1742-3; and is called by the natives *Ea-to-oobe*: it is about 16 miles from north to south, and in the widest part about 8 miles from east to west. The skirts are chiefly laid out in plantations, the south-west and north-west sides especially. The interior parts are but little cultivated, though very capable of it; but this neglect adds greatly to the beauty of the island; for here are agreeably dispersed groves of cocoa-nuts and other trees, lawns covered with thick grass, here and there plantations and paths leading to every part of the island, in such beautiful disorder, as greatly to enliven the prospect. The hills are low; the air is delightful; but unfortunately water is denied to this charming spot. Yams, with other roots, bananas, and bread-fruit, are the principle articles of food; but the latter appeared to be scarce. Here is the pepper-tree, or *ava-ava*, with which they make an intoxicating liquor, in the same disgusting manner as is practised in the Society Islands. Here are several odoriferous trees and shrubs,